Lecture Notes in Physics

Edited by H. Araki, Kyoto, J. Ehlers, München, K. Hepp, Zürich
R. Kippenhahn, München, D. Ruelle, Bures-sur-Yvette
H. A. Weidenmüller, Heidelberg, J. Wess, Karlsruhe and J. Zittartz, Köln
Managing Editor: W. Beiglböck

324

P. Exner P. Šeba (Eds.)

Applications of Self-Adjoint Extensions in Quantum Physics

Proceedings of a Conference Held at the Laboratory of Theoretical Physics, JINR
Dubna, USSR, September 29 – October 1, 1987

Springer-Verlag
Berlin Heidelberg GmbH

Editors

Pavel Exner
Petr Šeba
Laboratory of Theoretical Physics, JINR
141980 Dubna, USSR

ISBN 978-3-662-13762-8 ISBN 978-3-540-46104-3 (eBook)
DOI 10.1007/978-3-540-46104-3

Originally published by Springer-Verlag Berlin Heidelberg New York in 1989
Softcover reprint of the hardcover 1st edition 1989

2158/3140-543210

PREFACE

The idea of contact interaction first appeared in the formative years of quantum mechanics, when E.Fermi formulated in 1934 his famous deuteron model. Other applications in different fields, such as atomic and molecular physics, solid-state physics and scattering theory followed. The mathematical nature of contact interactions remained unclear, however, for a long time. It took thirty years from the publication of the original Fermi paper before a treatment of zero-range potentials by Berezin and Faddeev appeared. In a not very explicit way, this paper suggested the association of self-adjoint extensions of a suitable symmetric operator with the formal Schrödinger operator describing the point interactions. This connection is very important. The theory of self-adjoint extensions developed by J. von Neumann has been a standard part of functional analysis for more than half a century. Until recently, however, it had few physical applications apart from simple textbook examples, such as Schrödinger particles on a halfline.

The idea of using self-adjoint extensions to construct various contact-interaction models was recognized gradually as a very fruitful one; in particular, because the models obtained in this way are usually exactly solvable. In the last few years, we have witnessed a rapid development and many new applications of this method. The conference whose main results are summarized in this volume represented an attempt to bring together people who contributed to this development.

There are two schools of thought to which most of the people working in the field belong. The first of them is represented by the names of S.Albeverio, R.Høegh-Krohn and their collaborators. It is the more widely known approach and needs no introduction here, especially with the publication of the monograph summarizing their results in Springer's "Texts and Monographs in Physics" series.

Alternatively, there is a Soviet school resulting from the long mathematical tradition which can be traced back to the work of Krein and Vishik. In addition, many physical considerations have been summarized in the book of Demkov and Ostrovsky, which had, however, little impact outside the USSR. During recent years, the two traditional lines have been combined in Leningrad, where the group of B.Pavlov, himself a disciple of M.Krein, has obtained many remarkable results.

It is worth mentioning that the Leningrad group also contributed to the mathematical development of the self-adjoint extension theory. They devised a method which generalizes the original von Neumann scheme, making it applicable to non-densely defined operators as well. Since it is difficult to learn the method from the original sources even for Russian-reading people, we have asked Professor Pavlov to prepare a short introductory paper for this volume, in which the basic features will be outlined.

The Dubna conference succeeded in establishing for the first time contacts between representatives of the two schools. It proved to be useful, and we hope that the exchange of ideas has stimulated further progress in the field. The proceedings contain most of the main lectures and seminars presented at the conference. We have grouped them according to their physical content. Their common feature is that all of them use self-adjoint extensions to construct the quantities of physical interest; it makes the obtained models elegant and exactly solvable in most cases.

P. Exner

P. Šeba

Dubna, February 1988

Last Homage to a Great Scientist

The sad news about the untimely death of Raphael Høegh-Krohn reached us as this volume was nearing completion. It is painful to realize that a man so full of life and so rich in ideas has left us forever. Only a few months ago he participated in our conference, chairing its opening session and making a deep, personal contribution to numerous discussions.

During the past two decades, Raphael has contributed to the development of mathematical physics in many ways. One of his most outstanding achievements was the introduction, in collaboration with S.Albeverio and others, of a new approach to systems with contact interactions. It is a sad coincidence that a monograph summarizing the results of this approach appears just at this time.

The Dubna conference was devoted to a further development of these ideas and to the application of them to a wider class of physically interesting models. We therefore wish to dedicate the present volume to Raphael's memory. In our opinion, it is the best way to honour his scientific and human merits.

The Editors

TABLE OF CONTENTS

GENERAL CONSIDERATIONS

ZERO-RANGE INTERACTIONS WITH AN INTERNAL STRUCTURE

B.S.Pavlov

Department of Mathematical and Computational Physics, Institute for Physics, Leningrad State University, St.Peterhoff
198904 Leningrad, USSR

The purpose of this paper is to outline a general scheme for construction of zero-range interactions with an internal structure. This construction can be applied to various situations of physical interest - see [1-3]. Here we concentrate on the mathematical aspects of the problem.

A zero-range interaction with an internal structure is usually constructed as a self-adjoint extension of a direct sum of operators ; typically one of them is a differential operator and the other is an abstract one. The standard von Neumann theory works well for differential operators, though it is not very suitable for concrete calculations. A serious difficulty arises, however, if the abstract operator is not densely defined. The purpose of our construction is twofold :

(i) to reformulate the standard theory in terms of boundary conditions making it better adapted to practical calculations,

(ii) what is more important, to generalize it in such a way that it can handle also non-densely defined operators. Notice that it is necessary once we want to treat finite-dimensional Hilbert spaces.

The scheme of the construction described here has been published first in Ref.1 and discussed further in [2-5] ; see also the literature in [5].

1. The self-adjoint extensions

Let A be a self-adjoint operator in a Hilbert space $\mathcal{H}$ and let $U = (A-iI)(A+iI)^{-1}$ be its Cayley transform. Let $\mathcal{N}_i$ be some minimal generating subspace of A , i.e.,

(i) $\overline{\bigvee_l U^l \mathcal{N}_i} = \mathcal{H}$,

(ii) $\mathcal{N}_i \cap U\mathcal{N}_i = \{0\}$.

If $\dim \mathcal{H}_i < \infty$, the last condition is equivalent to

(iii) $\cos(\mathcal{H}_i, U\mathcal{H}_i) \equiv \sup\left\{ \frac{(u,v)}{\|u\|\|v\|} : u \in \mathcal{H}_i, \; v \in U\mathcal{H}_i \right\} < 1$,

which is stronger in general.

For further purpose, we denote $\mathcal{H}_{-i} \equiv U\mathcal{H}_i$. It is useful also to define analogous subspaces $\mathcal{H}_\lambda, \mathcal{H}_{\bar{\lambda}}$ for any complex λ outside $\sigma(A)$; they are related by $\mathcal{H}_{\bar{\lambda}} = U_\lambda \mathcal{H}_\lambda$ with $U_\lambda = (A-\lambda I)(A-\bar{\lambda} I)^{-1}$.

Now we are going to restrict the operator A . We introduce the following linear set in the domain D of A ,

$$D_0 = (A-iI)^{-1}\mathcal{H}_i^\perp = (A+iI)^{-1}\mathcal{H}_{-i}^\perp ,$$

and restrict the operator A to $A_0 \equiv A \upharpoonright D_0$. Our aim is to construct all self-adjoint extensions A_Γ of A_0 . In distinction to the standard theory, we do not require A_0 to be densely defined. Then the adjoint A_0^* may not exist, and consequently, the definition of deficiency subspaces must be used in the form

$$\mathcal{K}_{\pm i} = [\mathrm{Ran}(A_0 \mp iI)]^\perp .$$

The following assertion is valid :

<u>Lemma 1</u> : (i) The deficiency subspaces $\mathcal{K}_{\pm i}$ of A_0 coincide with $\mathcal{H}_{\pm i}$.

(ii) The operator A_0 is densely defined iff $D \quad _{\pm i} = 0$.

The analogous assertion can be proven for the subspaces $\mathcal{H}_\lambda$ corresponding to any $\lambda \in \mathbb{C} \setminus \sigma(A)$.

<u>Proof</u> : (i) follows directly from the definition of D_0 and (ii) is a consequence of the relation

$$\langle (A \mp iI)u_0, \theta_{\pm i} \rangle = \langle u_0, (A \pm iI)\theta_{\pm i} \rangle = 0$$

which is valid for every $u_0 \in D_0$, $\theta_{\pm i} \in \mathcal{H}_{\pm i} \cap D$ since $A \pm iI$ maps D_0 into $\mathcal{H}_{\pm i}^\perp$.

The construction of the extensions A_Γ proceeds in a usual way. We restrict the Cayley transform U to $U_0 \equiv U \upharpoonright \mathcal{H}_{-i}^\perp$ and construct all isometries $\hat{U}_\Gamma : \mathcal{H}_{-i} \to \mathcal{H}_{+i}$. Combining U_0 and $\hat{U}_\Gamma$, we get a unitary operator whose Cayley pull-back is the sought extension A_Γ .

Next we need a technical lemma :

Lemma 2 : If the condition (iii) is fulfilled and $\{\theta_s\}$ is a Riesz basis (see [6]) in $\mathcal{H}_i$, then the vectors $\{W_s^+ , W_s^-\}$ defined by

$$W_s^+ = \frac{U^*+I}{2}\,\theta_s \quad , \quad W_s^- = \frac{U^*-I}{2}\,\theta_s$$

form a Riesz basis in the subspace $\hat{\mathcal{H}} = \mathcal{H}_i + \mathcal{H}_{-i}$.

Proof is obvious.

Hence any vector $u \in \hat{\mathcal{H}}$ can be decomposed as

$$u = \sum_s (\xi_s^+ W_s^+ + \xi_s^- W_s^-) \quad , \quad \xi_s^{\pm} \in \mathbf{C} \quad .$$

Introducing the vectors $\xi^{\pm} = \sum_s \xi_s^{\pm}\theta_s \in \mathcal{H}_i$, we can rewrite this decomposition as

$$u = A(A-iI)^{-1}\xi^+ + (A-iI)^{-1}\xi^- \quad .$$

Using the basis $\{W_s^+, W_s^-\}$, we define further the linear operator $\hat{A} : \hat{\mathcal{H}} \to \hat{\mathcal{H}}$ by the relations

$$\hat{A}W_s^+ = -W_s^- \quad , \quad \hat{A}W_s^- = W_s^+ \quad .$$

If A_0 is densely defined, then $\hat{A} = A_0^* \upharpoonright \hat{\mathcal{H}}$. It suggests that we can use this operator for construction of the extensions even if A_0^* does not exist.

Remark : In most of applications, in particular, those mentioned above [1-5] we use formally the symbol A_0^* . In the case of non-densely defined operators, it should be understood as described here.

Lemma 3 : Under the hypotheses of Lemma 2, the following formula is valid

$$\langle Au,v\rangle - \langle u,Av\rangle = \sum_s \{\xi_s^-(u)\overline{\xi_s^+(v)} - \xi_s^+(u)\overline{\xi_s^-(v)}\} =$$

$$= \langle \xi^-(u),\xi^+(v)\rangle - \langle \xi^+(u),\xi^-(v)\rangle \quad ,$$

where u,v are arbitrary vectors of $\hat{\mathcal{H}}$ with the decompositions

$$u = \sum_s \{\xi_s^+(u)W_s^+ + \xi_s^-(u)W_s^-\} = A(A-iI)^{-1}\xi^+(u) + (A-iI)^{-1}\xi^-(v) \quad ,$$

$$v = \sum_s \{\xi_s^+(v) W_s^+ + \xi_s^-(v) W_s^-\} = A(A-iI)^{-1}\xi^+(v) + (A-iI)^{-1}\xi^-(v) \quad .$$

Proof : It follows from the definition of the basis $\{W_s^+, W_s^-\}$ that

$$\langle W_s^+, W_t^+\rangle - \langle W_s^-, W_t^-\rangle = \langle \Theta_s, \Theta_t\rangle \quad ,$$
$$\langle W_s^+, W_t^-\rangle - \langle W_s^-, W_t^+\rangle = 0$$

for all s,t . The result is now obtained by a straightforward calculation.

Lemma 3 provides us with the universal symplectic boundary form

$$J(u,v) \equiv \langle Au, v\rangle - \langle u, Av\rangle = \langle \xi^-(u), \xi^+(v)\rangle - \langle \xi^+(u), \xi^-(v)\rangle \quad ,$$

where the rhs gives the expression of $J(u,v)$ in terms of the vectors $\xi^{\pm}$. The latter play here the role of usual boundary values known from the theory of differential operators.

If A is an orthogonal sum of self-adjoint operators,

$$A = A_0 \oplus A_1$$

on the Hilbert space $\mathcal{H} = \mathcal{H}_0 \oplus \mathcal{H}_1$, then its boundary form equals to the sum of the boundary forms of its parts,

$$J(u,v) = J_0(u_0,v_0) + J_1(u_1,v_1) \quad ,$$

where $u = (u_0,u_1)$ and $v = (v_0,v_1)$.

Now we are going to formulate the main result :

Theorem 1 : Under the assumptions (i) and (ii), all self-adjoint extensions of the operator A_0 are determined by Lagrange hyperplanes in $\hat{\mathcal{H}}$ on which the form J vanishes. Each of the hyperplanes can be specified by the boundary conditions

$$\xi^- = \Gamma \xi^+ \quad , \tag{1}$$

where Γ is a self-adjoint operator, $\Gamma = \Gamma^*$, in $\mathcal{H}_i$.

Remark : In order to describe all the self-adjoint extensions, one can start with a fixed hyperplane. All other Lagrange hyperplanes can be obtained from the initial one by J-unitary transformations.

Proof : What we need is to relate the boundary conditions (1) to the isometries $\hat{U}_\Gamma : \mathcal{H}_{-i} \to \mathcal{H}_i$. We define $\hat{U}_\Gamma$ as the Cayley transform of the operator $\hat{A}_\Gamma$ obtained as the restriction of $\hat{A}$ to the linear manifold determined by the boundary conditions (1). Using the relations

$$(\hat{A}-iI)\sum_s (\xi_s^+ w_s^+ + \xi_s^- w_s^-) = \sum_s ((\xi_s^- - i\xi_s^+)w_s^+ - (\xi_s^+ + i\xi_s^-)w_s^-) =$$

$$= A(A-iI)^{-1}(\xi^- - i\xi^+) - (A-iI)^{-1}(\xi^+ + i\xi^-) = (\Gamma - iI)\xi^+ \equiv v$$

and

$$(A+iI)\sum_s (\xi_s^+ w_s^+ + \xi_s^- w_s^-) = A(A-iI)^{-1}(\xi^- + i\xi^-) - (A-iI)^{-1}(\xi^+ - i\xi^-) =$$

$$= (A+iI)(A-iI)^{-1}(\Gamma + iI)\xi^+ \equiv w \quad ,$$

we see that the Cayley transform $\hat{U}_\Gamma$ of $\hat{A}_\Gamma$ maps $\mathcal{H}_{-i}$ into $\mathcal{H}_i$ according to $v = \hat{U}_\Gamma w$, i.e.,

$$\hat{U}_\Gamma = (\Gamma - iI)(\Gamma + iI)^{-1}(A + iI)^{-1}(A - iI) \quad . \tag{2}$$

Hence it differs from U by the factor $(\Gamma - iI)(\Gamma + iI)^{-1}$ only, and the latter is a unitary operator in $\mathcal{H}_i$. However, if Γ runs through the set of all self-adjoint operators on $\mathcal{H}_i$, the formula (2) yields a parametrization of all isometries $\mathcal{H}_{-i} \to \mathcal{H}_i$. The only exception is the original isometry $\hat{U}_\Gamma = U$. Using now the Cayley pull-back of $\hat{U}_\Gamma$, we get the sought self-adjoint extension $\hat{A}_\Gamma$. The only exception is again the original operator A to which the boundary condition $\xi^+ = 0$ corresponds.

Corollary 1 : The domain D_Γ of the extension $\hat{A}_\Gamma$ consists of the vectors

$$u = u_0 + A(A-iI)^{-1}\xi^+ + (A-iI)^{-1}\xi^- = u_0 + [A(A-iI)^{-1} + (A-iI)^{-1}]\xi^+ , \tag{3}$$

where $u_0 \in D_0$ and ξ^+ runs through $\mathcal{H}_i$.

It means that the domain D_Γ of $\hat{A}_\Gamma$ can be represented in the form

$$D = D_0 + \hat{\mathcal{H}}_\Gamma , \tag{4}$$

where $\hat{\mathcal{H}}_\Gamma$ is the Lagrange hyperplane in $\hat{\mathcal{H}}$ specified by the boundary condition (1). To get this representation, we employ the relations between A and U ,

$$D_\Gamma = (\hat{A}_\Gamma+iI)^{-1}\mathcal{H} = \frac{1}{2i}(I-\hat{U}_\Gamma)\mathcal{H}_{-i}^{\perp} + \frac{1}{2i}(I-\hat{U}_\Gamma)\mathcal{H}_{-i} =$$

$$= \frac{1}{2i}(I-U)\mathcal{H}_{-i}^{\perp} + \frac{1}{2i}(I-\hat{U}_\Gamma)\mathcal{H}_{-i} = D_0 + \hat{\mathcal{N}}_\Gamma .$$

The last decomposition is a direct sum if $\mathcal{N}_i \subset D$ and the operator $PAP+\Gamma$ is invertible, where P is the projection to $\mathcal{N}_i$. In fact, in this case D_0 and $\hat{\mathcal{N}}_\Gamma$ are linearly independent for

$$(A-iI)^{-1}h + A(A-iI)^{-1}\xi^+ + \Gamma(A-iI)^{-1}\xi^+ = 0$$

if and only if $P(A+\Gamma)\xi^+ = 0$, i.e., $\xi^+ = 0$ and $h = 0$. Then both the components in the decomposition $u = u_0 + h$, where $u_0 \in D_0$ and $h \in \hat{\mathcal{N}}_\Gamma$, are uniquely determined,

$$h = [A(A-iI)^{-1} + \Gamma(A-iI)^{-1}]\xi^+ ,$$

$$\xi^+ = (PAP+\Gamma)^{-1}P(A-iI)u .$$

In conclusion, let us illustrate the construction described above.

<u>Example</u> : Consider the simplest situation involving a non-densely defined operator in a finite-dimensional Hilbert space $\mathcal{H}$, $\dim\mathcal{H} = n$. Let $A = \sum_s \alpha_s e_s \langle \cdot , e_s \rangle$ be a self-adjoint operator in $\mathcal{H}$ with a simple spectrum, $\alpha_s \neq \alpha_t$ for $s \neq t$. Let $\mathcal{N}_i = \{g\}$ be the subspace spanned by a vector $g = \sum_s g_s e_e$. Then $\mathcal{N}_{-i} = \{g'\}$, where $g' =$ $= \sum_s (\alpha_s+i)(\alpha_s-i)^{-1} g_s e_s$ and the elements

$$W^+ = A(A-iI)^{-1}g , \quad W^- = (A-iI)^{-1}g$$

form a basis in $\hat{\mathcal{N}} = \mathcal{N}_i + \mathcal{N}_{-i}$. The Lagrange line specified by the boundary condition $\xi^- = \Gamma\xi^+$ is spanned by the vector

$$\xi^+ \sum_s (\alpha_s+\Gamma)(\alpha_s-i)^{-1} g_s e_s ,$$

and the decomposition (4) is a direct sum if $\langle Ag,g \rangle + \Gamma \neq 0$. We have

$$D_\Gamma = \{ x : x = \sum_s (\alpha_s-i)^{-1}[h_s - \langle h,g \rangle g_s + \xi^+(\alpha_s+\Gamma)g_s]e_s , \ h \in \mathcal{H} \} ,$$

$$Ax = \sum_s \alpha_s(\alpha_s - i)^{-1}\left[h_s - \langle h,g\rangle g_s + \xi^+(\alpha_s+\Gamma)g_s\right]e_s \quad ,$$

$$A_\Gamma x = \sum_s (\alpha_s - i)^{-1}\left[\alpha_s(h_s - \langle h,g\rangle g_s) + \xi^+(\Gamma\alpha_s - 1)g_s\right]e_s$$

so A and A_Γ differ by a rank-one operator only,

$$(A_\Gamma - A)x = -\xi^+(x)\,(A + iI)g \quad ,$$

where $\xi^+(x) = \left[\langle Ag,g\rangle + \Gamma\right]^{-1}\langle (A-iI)x,g\rangle$.

2. Krein formula

In this section, we are going to derive Krein's formula for resolvents. Let $A_\Gamma = A_0 + \hat{A}_\Gamma$ be one of the self-adjoint extensions constructed in the preceding section. To obtain its resolvent, we have to solve the equation

$$(A_\Gamma - \lambda I)u = f \tag{5}$$

for all $f \in \mathcal{H}$. First we prove

Lemma 4 : The deficiency subspaces $\mathcal{N}_i$ and $\mathcal{N}_{\bar\lambda}$ corresponding to i and $\bar\lambda$, respectively, are related by

$$\mathcal{N}_{\bar\lambda} = \frac{A+iI}{A-iI}\mathcal{N}_i \quad . \tag{6}$$

Proof : The result follows from the relation (in an obvious notation)

$$0 = \langle (A - \bar\lambda I)D_0, \mathcal{N}_\lambda\rangle = \langle (A - iI)D_0, (A + iI)^{-1}(A - \lambda I)\mathcal{N}_\lambda\rangle .$$

Hence $(A+iI)^{-1}(A-\lambda I)\mathcal{N}_\lambda \perp (A-iI)D_0$ and Lemma 1 applies.

The relation (6) defines an isomorphism of $\mathcal{N}_i$ with $\mathcal{N}_{\bar\lambda}$ which ascribes to every deficiency vector $\theta_i \in \mathcal{N}_i$ a deficiency vector $\theta_\lambda \in \mathcal{N}_{\bar\lambda}$.

Lemma 5 : The deficiency vectors θ_λ and θ_i are related by

$$\theta_\lambda = \frac{1}{A-iI}(I-P)\frac{I+A}{A-\lambda I}\theta_i + \frac{A}{A-iI}\theta_i + \frac{1}{A-iI}P\frac{I+A}{A-\lambda I}\theta_i \quad , \tag{7}$$

where P is the orthogonal projection onto $\mathcal{N}_i$. The first term on the rhs belongs to D_0 , while the remaining ones to $\hat{\mathcal{N}}$.

Proof is based on the relations (6) and

$$A + iI = (A-iI)^{-1}\left[I + \lambda A + A(A-\lambda I)\right]$$

together with

$$(A-iI)^{-1}(I-P)(A-\lambda I)^{-1}(I+\lambda A)\,\theta_i \in D_0 \quad .$$

The formula (7) shows that θ_λ can be expressed in the form (3) if we choose

$$\xi^+ = \theta_i \quad ,$$

$$\xi^- = P(A-\lambda I)^{-1}(I+\lambda A)\,\theta_i = Q(\lambda)\,\theta_i \quad .$$

The function $Q(\lambda)$ whose values are operators in $\mathcal{H}_i$ is a very important object in our theory ; it is called Schwartz integral of the operator A associated with the deficiency subspace $\mathcal{H}_i$:

$$Q(\lambda) = \int \frac{1+\lambda s}{s-\lambda}\, d(PE_sP) \quad , \qquad (8)$$

where E_s is the spectral measure of A . It is clear that $Q(.)$ is an operator-valued R-function, i.e., its imaginary part is positive in the upper halfplane. Since $\mathcal{H}_i$ is supposed to be a generating subspace we see that $Q(\lambda)$ contains all the spectral information about A . Hence it plays a crucial role in construction of the resolvent of A .

Theorem 2 : The resolvent $(A_\Gamma-\lambda I)^{-1}$ is given by the Krein's formula

$$(A_\Gamma-\lambda I)^{-1} = (A-\lambda I)^{-1} + (A+iI)P\left[\Gamma - Q(\lambda)\right]^{-1} P(A-iI)(A-\lambda I)^{-1} \quad .$$

Proof : According to Corollary 1, we look for a solution to (5) in the form (3) with $u_0 = (A-iI)^{-1}v^\perp$, where ξ^+ and ξ^- are related by the condition (1) and $v^\perp \in \mathcal{H}_i$. Using the identities

$$\hat{A}A(A-iI)^{-1}\xi^+ = -(A-iI)^{-1}\xi^- \quad ,$$

$$A(A-iI)^{-1}\xi^- = A(A-iI)^{-1}\xi^+$$

which follow from the definition of $\hat{A}$, we rewrite the eq.(5) in the form

$$(A_{\Gamma}-\lambda I)u = (A-\lambda I)(A-iI)^{-1}v^{\perp} + (A-\lambda I)(A-iI)^{-1}\xi^{-} - (I+ A)(A-iI)^{-1}\xi^{+} = \tag{9}$$

$$= f \quad .$$

Applying the projection P to the both sides and multiplying by the bounded operator $(A-iI)(A-\lambda I)^{-1}$, we get

$$\Gamma\xi^{+} = \xi^{-} = P(A-\lambda I)^{-1}(I+\lambda A)\xi^{+} + P(A-iI)(A-\lambda I)^{-1}f \quad ,$$

and therefore

$$\xi^{+} = [\Gamma - Q(\lambda)]^{-1} P (A-iI)(A-\lambda I)^{-1}f \quad .$$

Substituting this to (9) we find $v^{\perp}$ so (3) yields the sought vector u .

References

1 B.S.Pavlov, Teor.Mat.Fiz.59 (1984),159-169.

2 Yu.A.Kuperin, K.A.Makarov, B.S.Pavlov, Teor.Mat.Fiz.69 (1986), 100-114.

3 V.M.Adamian, B.S.Pavlov, Leningrad Branch of Steklov Institute Zapisky 149 (1986),7-23.

4 Yu.A.Kuperin, K.A.Makarov, S.P.Merkuriev, A.K.Motovilov, B.S.Pavlov : Quantum scattering theory with energy-dependent potentials, part II, Vilnius 1986 ; pp.28-73.

5 B.S.Pavlov, Sov.Math.Uspekhi 42 (1987), No.6, 99-131.

6 I.C.Gohberg, M.G.Krein : Introduction to the Theory of Linear Non-selfadjoint Operators, Nauka, Moscow 1965 (in Russian).

7 M.G.Krein, Acad.Sci.USSR Doklady 52 (1946),657-660.

EVOLUTION EQUATIONS
AND
SELFADJOINT EXTENSIONS

Hagen Neidhardt+)

Laboratory of Theoretical Physics
Joint Institute for Nuclear Research
141980 Dubna, USSR

1. Introduction

Let $\mathfrak{H}$ be a separable Hilbert space with scalar product $\langle .,. \rangle$ and norm $\|.\|$. We assume that we have a family $\{H(t)\}_{t\in\mathbb{R}^1}$ of selfadjoint operators on $\mathfrak{H}$. For instance, this can be a family of Schrödinger operators $\{-\Delta + q(t)\}_{t\in\mathbb{R}^1}$ on $L^2(\mathbb{R}^n)$ which naturally arises from a time-dependent potential $q(.,t)$. With $\{H(t)\}_{t\in\mathbb{R}^1}$ we associate an evolution equation

$$i\frac{du(t)}{dt} = H(t)u(t), \tag{1.1}$$

$$u(t)|_{t=s} = x, \tag{1.2}$$

$x \in \mathfrak{H}$, $t,s \in \mathbb{R}^1$. The problem is to find the solution operator of this evolution equation, i.e. a strongly continuous family $\{U(t,s)\}_{t,s\in\mathbb{R}^1}$ of unitary operators on $\mathfrak{H}$ such that in some sense $u(t) = U(t,s)x$ is a solution of (1.1) and (1.2). To attack this problem there are several possibilities.

The first investigations of this problem were carried out by T.Kato [1] who returned to it several times [2-6]. But there is a lot of other authors who were interested in the same problem, among them H.Tanabe, P.E.Sobolevskij, J.A.Goldstein, S.G.Krein, J.R.Dorroh, A.Yagi etc.. The methods used by these authors are similar to those of the theory of ordinary differential equations.The main task is to find suitable equations, which approximate the original equation and which are easy to solve, and to show that the arising approximation solution operators converge to that of the original problem.

+) On leave of absence from Karl-Weierstraß-Institut für Mathematik, Akademie der Wissenschaften der DDR, 1086 Berlin, DDR

An application of the abstract results to the time-dependent Schrödinger operator $\{-\Delta + q(t)\}_{t\in\mathbb{R}^1}$ can be found in [7,8].

In the following we use an approach to this problem which is quite different from the mentioned one and which has been proposed in [9,10]. This approach goes back to ideas, which can be found by O.A.Ladyženskaja [11,12], J.L.Lions [13,14,15], G.DaPrato [16,17], M.Iannelli [18], G.DaPrato and P.Grisvard [19] and L.Paquet [20]. In application to the present Hilbert space situation the method can be described as follows.

Let z be a complex number, $\mathrm{Im}(z) \neq 0$, such that for every $x \in \mathfrak{h}$ the map

$$(H(.) - z)^{-1} x\colon \mathbb{R}^1 \longmapsto \mathfrak{h} \tag{1.3}$$

is measurable. By $L^2(\mathbb{R}^1, \mathfrak{h})$ we denote the Hilbert space of all measurable $\mathfrak{h}$-valued functions on $\mathbb{R}^1$ such that $f(.)\|f(.)\|$ is Bochner integrable. The scalar product $(.,.)$ of $L^2(\mathbb{R}^1,\mathfrak{h})$ is given by

$$(f,g) = \int_{\mathbb{R}^1} \langle f(t), g(t)\rangle dt \tag{1.4}$$

$f,g \in L^2(\mathbb{R}^1,\mathfrak{h})$. With the family $\{H(t)\}_{t\in\mathbb{R}^1}$ we associate a selfadjoint operator H on $L^2(\mathbb{R}^1,\mathfrak{h})$ defined by

$$\mathrm{dom}(H) = \Big\{ f \in L^2(\mathbb{R}^1,\mathfrak{h})\colon f(t) \in \mathrm{dom}(H(t)) \text{ a.e. } t\in\mathbb{R}^1, \\ H(t)f(t) \in L^2(\mathbb{R}^1,\mathfrak{h}) \Big\}, \tag{1.5}$$

$$(Hf)(t) = H(t)f(t), \tag{1.6}$$

$f \in \mathrm{dom}(H)$. The selfadjointness of H is guaranteed by the measurability condition (1.3).

Further we consider the shift group $e^{-i\sigma K_0}$, $\sigma \in \mathbb{R}^1$, on $L^2(\mathbb{R}^1,\mathfrak{h})$ given by

$$(e^{-i\sigma K_0} f)(t) = f(t - \sigma), \tag{1.7}$$

$t \in \mathbb{R}^1$, f $L^2(\mathbb{R}^1,\mathfrak{h})$. It is easy to show that $K_0 = -i\frac{d}{dt}$, where $-i\frac{d}{dt}$ is defined via the Fourier transform as the usual multiplication operator induced by the independent variable t.

We assume that the set $\mathrm{dom}(\widetilde{K})$,

$$\mathrm{dom}(\tilde{K}) = \mathrm{dom}(K_0) \cap \mathrm{dom}(H) \tag{1.8}$$

is dense in $L^2(\mathbb{R}^1, \mathfrak{h})$. Obviously, this condition implies some restrictions on $\{H(t)\}_{t\in\mathbb{R}^1}$ but they are not very hard. Then we introduce the operator $\tilde{K}$,

$$\tilde{K}f = K_0 f + Hf, \tag{1.9}$$

$f \in \mathrm{dom}(\tilde{K})$. The operator $\tilde{K}$ is densely defined and symmetric but in general not closed. Moreover, denoting by $M(\phi)$, $\phi \in C_0^1(\mathbb{R}^1)$, the bounded multiplication operator induced by $\phi(.)$, i.e.

$$(M(\phi)f)(t) = \phi(t)f(t), \tag{1.10}$$

where $C_0^1(\mathbb{R}^1)$ is the set of all continuously differentiable functions $\phi: \mathbb{R}^1 \longrightarrow \mathbb{C}^1$ with compact support, we find

$$M(\phi)\mathrm{dom}(\tilde{K}) \subseteq \mathrm{dom}(\tilde{K}) \tag{1.11}$$

and

$$\tilde{K}M(\phi)f - M(\phi)\tilde{K}f = -iM(\dot{\phi})f, \tag{1.12}$$

$\dot{\phi} = \frac{d\phi}{dt}$, for every $f \in \mathrm{dom}(\tilde{K})$ and every $\phi \in C_0^1(\mathbb{R}^1)$. Let $C(\mathbb{R}^1, \mathfrak{h})$ be the set of all continuous functions $f(.): \mathbb{R}^1 \longrightarrow \mathfrak{h}$. Then, in addition we have

$$\mathrm{dom}(\tilde{K}) \subseteq L^2(\mathbb{R}^1, \mathfrak{h}) \cap C(\mathbb{R}^1, \mathfrak{h}). \tag{1.13}$$

This inclusion follows from $\mathrm{dom}(\tilde{K}) \subseteq \mathrm{dom}(K_0) \subseteq L^2(\mathbb{R}^1, \mathfrak{h}) \cap$ $\cap\, C(\mathbb{R}^1, \mathfrak{h})$.

Denoting by K the closure of $\tilde{K}$ it is not hard to show [10] that K fulfils the conditions (1.11) - (1.13), too. This means we have

$$M(\phi)\mathrm{dom}(K) \subseteq \mathrm{dom}(K) \tag{1.14}$$

and

$$KM(\phi)f - M(\phi)Kf = -iM(\dot{\phi})f \tag{1.15}$$

for every $f \in \mathrm{dom}(K)$ and every $\phi \in C_0^1(\mathbb{R}^1)$ as well as

$$\mathrm{dom}(K) \subseteq L^2(\mathbb{R}^1,\mathfrak{h}) \cap C(\mathbb{R}^1,\mathfrak{h}). \qquad (1.16)$$

The last fact needs some more efforts than the other two.

Assuming now for a moment that K is selfadjoint and that the conditions (1.14) and (1.15) are fulfilled, then in accordance with J.S.Howland [21] there is a measurable family $\{U(t,s)\}_{t,s\in\mathbb{R}^1}$ of unitary operators such that the representation

$$(e^{-i\sigma K} f)(t) = U(t,t-\sigma)f(t-\sigma), \qquad (1.17)$$

$f \in L^2(\mathbb{R}^1,\mathfrak{h})$, is valid. Furthermore, if (1.16) is satisfied and the set

$$(\mathrm{dom}(K))_t = \{x \in \mathfrak{h} : \exists f \in \mathrm{dom}(K),\ x = f(t)\}, \qquad (1.18)$$

$t \in \mathbb{R}^1$, is dense in $\mathfrak{h}$ for every $t \in \mathbb{R}^1$, then $\{U(t,s)\}_{t,s\in\mathbb{R}^1}$ can be continuously chosen [22].

As has been pointed out in [10] this continuous two-parameter family $\{U(t,s)\}_{t,s\in\mathbb{R}^1}$ of unitary operators can be regarded as the first candidate for a solution operator of the evolution equation (1.1) and (1.2). In [10] additional conditions were made which really allow to verify the usual properties of a solution operator for $\{U(t,s)\}_{t,s\in\mathbb{R}^1}$. Several applications of this approach to evolution equations were given in [23].

If now K is not selfadjoint, then naturally the problem arises to extend K to a selfadjoint operator $\hat{K}$ preserving the conditions (1.14) - (1.16) and (1.18). But simple examples show that that is not always possible. Hence, we need conditions which guarantee the existence of such an extension $\hat{K}$ of K. To find such conditions will be the aim of the present note.

As the first step on this way we concentrate our attention on the conditions (1.14) and (1.15) and forget (1.16) and (1.18). Therefore, the problem will be now to find conditions allowing the existence of symmetric, maximal symmetric and selfadjoint extensions of K obeying (1.14) and (1,15).

Besides, a further motivation of this problem comes from the so-called Weyl commutation-relation [24]. Let Q be the selfadjoint multiplication operator on $L^2(\mathbb{R}^1,\mathfrak{h})$ induced by the variable t, i.e.

$$(Qf)(t) = tf(t), \tag{1.19}$$

$f \in \mathrm{dom}(Q) = \{ f \in L^2(\mathbb{R}^1, \mathfrak{h}) : tf(t) \in L^2(\mathbb{R}^1, \mathfrak{h}) \}$. If K is selfadjoint and obeys (1.14) and (1.15), then using the considerations of J.S.Howland [21] $\{K,Q\}$ forms a Weyl pair, i.e.

$$e^{-i\sigma K}\, e^{-i\tau Q} = e^{i\sigma\tau}\, e^{-i\tau Q}\, e^{-i\sigma K}, \tag{1.20}$$

$\sigma, \tau \in \mathbb{R}^1$. If K is not selfadjoint but obeys (1.14) and (1.15), then $\{K,Q\}$ forms in some sense an incomplete Weyl pair. Hence, to solve the above proposed problem means to complete $\{K,Q\}$ to a Weyl pair $\{\hat{K},Q\}$, $\hat{K} \supset K$. From this view-point the paper is related to that of B.Fuglede [25], N.S.Poulsen [26], P.E.T.Jørgensen [27], P.E.T.Jørgensen and P.S.Muhly [28], W.J.Phillips [29], K.Schmüdgen [30,31], G.Dorfmeister and J.Dorfmeister [32].

In the end we remark that unfortunately page restrictions make it impossible for us to present the proofs of the results. The reader, who is interested in proofs, must be refered to [33].

2. The associated representation

We introduce the set $AC(\mathbb{R}^1)$ of all locally absolutely continuous functions $\phi : \mathbb{R}^1 \mapsto \mathbb{C}^1$ such that ϕ and $\dot{\phi}$ belong to $L^\infty(\mathbb{R}^1)$.

Lemma 2.1. Let K be a symmetric operator on $L^2(\mathbb{R}^1, \mathfrak{h})$ obeying (1.14) and (1.15). Then, for every $\phi \in AC(\mathbb{R}^1)$ we have

$$M(\phi)\mathrm{dom}(K^*) \subseteq \mathrm{dom}(K^*) \tag{2.1}$$

and

$$K^*M(\phi)f - M(\phi)K^*f = -iM(\dot{\phi})f, \tag{2.2}$$

$f \in \mathrm{dom}(K^*)$.

The proof is obvious. Since every symmetric extension of K is a restriction of K^* to some domain $\mathcal{D}$, $\mathrm{dom}(K) \subseteq \mathcal{D} \subseteq \mathrm{dom}(K^*)$, the problem to obtain an extension of K preserving (1.14) and (1.15) reduces to the existence of a domain $\mathcal{D}$, $\mathrm{dom}(K) \subseteq \mathcal{D} \subseteq \mathrm{dom}(K^*)$ such that

$$M(\phi)\mathcal{D} \subseteq \mathcal{D} \tag{2.3}$$

is valid for every $\phi \in C_0^1(\mathbb{R}^1)$ and $\hat{K} = K \upharpoonright \mathcal{D}$ is symmetric.

By $\mathcal{R}_{-i}$ and $\mathcal{R}_i$ we denote the defect spaces

$$\mathcal{R}_{-i} = L^2(\mathbb{R}^1, \mathfrak{h}) \ominus \text{ima}(K + i) = \ker(K^* - i) \qquad (2.4)$$

and

$$\mathcal{R}_i = L^2(\mathbb{R}^1, \mathfrak{h}) \ominus \text{ima}(K - i) = \ker(K^* + i), \qquad (2.5)$$

where ima(.) is the range of an operator. The ordered pair $\{\dim(\mathcal{R}_{-i}), \dim(\mathcal{R}_i)\}$ is called the defect index of the operator K. It is well-known that there exists a selfadjoint extension of K if and only if $\dim(\mathcal{R}_{-i}) = \dim(\mathcal{R}_i)$.

Now the domain $\text{dom}(K^*)$ admits the decomposition

$$\text{dom}(K^*) = \text{dom}(K) \dotplus \mathcal{R}_{-i} \dotplus \mathcal{R}_i, \qquad (2.6)$$

where $X \dotplus Y$ means $X \cap Y = \{0\}$ [34]. Because of (2.6) we have

$$M(\phi)g_1 = f_1(\phi) + g_{11}(\phi) + g_{21}(\phi), \qquad (2.7)$$

$f_1(\phi) \in \text{dom}(K)$, $g_{11}(\phi) \in \mathcal{R}_{-i}$, $g_{21}(\phi) \in \mathcal{R}_i$, for every $g_1 \in \mathcal{R}_{-i}$ and $\phi \in AC(\mathbb{R}^1)$. Similarly, we get

$$M(\phi)g_2 = f_2(\phi) + g_{12}(\phi) + g_{22}(\phi), \qquad (2.8)$$

$f_2(\phi) \in \text{dom}(K)$, $g_{21}(\phi) \in \mathcal{R}_{-i}$, $g_{22}(\phi) \in \mathcal{R}_i$, for every $g_2 \in \mathcal{R}_i$ and $\phi \in AC(\mathbb{R}^1)$. Hence, we can introduce the linear operators $A_{ij}(\phi)$,

$$g_{ij}(\phi) = A_{ij}(\phi)g_j, \qquad (2.9)$$

$i,j = 1,2$, acting between the spaces $\mathcal{R}_{-i}$ and $\mathcal{R}_i$.

Further we say a sequence $\{\phi_n\}_{n=1}$ of elements of $AC(\mathbb{R}^1)$ converges pointwise to an element ϕ of $AC(\mathbb{R}^1)$ if

$$\sup_n \{ |\dot{\phi}_n|_{L^\infty(\mathbb{R}^1)} + |\phi_n|_{L^\infty(\mathbb{R}^1)} \} < +\infty, \qquad (2.10)$$

$$\lim_{n \to +\infty} \phi_n(0) = \phi(0), \qquad (2.11)$$

$$\lim_{n \to +\infty} \dot{\phi}_n(t) = \dot{\phi}(t) \qquad (2.12)$$

for a.e. $t \in \mathbb{R}^1$. It is not hard to see that (2.10) - (2.12) imply $\lim_{n \to +\infty} \phi_n(t) = \phi(t)$ for every $t \in \mathbb{R}^1$.

Moreover, we introduce the Hilbert space $\mathcal{K} = \mathcal{R}_{-i} \times \mathcal{R}_i$ with the scalar product

$$\left(\begin{pmatrix} f_1 \\ f_2 \end{pmatrix}, \begin{pmatrix} g_1 \\ g_2 \end{pmatrix}\right)_{\mathcal{K}} = (f_1, g_1) + (f_2, g_2), \tag{2.13}$$

$\begin{pmatrix} f_1 \\ f_2 \end{pmatrix} \in \mathcal{K}$, $\begin{pmatrix} g_1 \\ g_2 \end{pmatrix} \in \mathcal{K}$, $f_1, g_1 \in \mathcal{R}_{-i}$, $f_2, g_2 \in \mathcal{R}_i$ and the norm $\|.\|_{\mathcal{K}}$. Equipping the Hilbert space $\mathcal{K}$ with the bilinear form $[.,.]$,

$$[f,g] = (Jf, g)_{\mathcal{K}}, \tag{2.14}$$

$f, g \in \mathcal{K}$, where J is a bounded operator on $\mathcal{K}$ given by

$$J\begin{pmatrix} f_1 \\ f_2 \end{pmatrix} = \begin{pmatrix} f_1 \\ -f_2 \end{pmatrix}, \tag{2.15}$$

$\mathcal{K}$ transforms into a Krein space. An introduction to the theory of Krein spaces can be found in [35,36]. By means of the bilinear form $[.,.]$ it is possible to define the J-adjoint operator of a linear, densely defined operator on [35,36]. We denote the J-adjoint of an operator A on $\mathcal{K}$ by $A^{[*]}$. We have

$$A^{[*]} = JA^*J. \tag{2.16}$$

We introduce the map $AC(\mathbb{R}^1) \ni \phi \longrightarrow \mathcal{L}(\mathcal{K})$, where $\mathcal{L}(\mathcal{K})$ is the set of linear operators on $\mathcal{K}$ defined by

$$A(\phi)f = \begin{pmatrix} A_{11}(\phi) & A_{12}(\phi) \\ A_{21}(\phi) & A_{22}(\phi) \end{pmatrix} \begin{pmatrix} f_1 \\ f_2 \end{pmatrix}, \tag{2.17}$$

$f = \begin{pmatrix} f_1 \\ f_2 \end{pmatrix} \in \mathcal{K}$.

<u>Theorem 2.1.</u> The map $AC(\mathbb{R}^1) \ni \phi \longrightarrow \mathcal{L}(\mathcal{K})$ defined by (2.17) satisfies the following properties:

(i) For every $\phi \in AC(\mathbb{R}^1)$, $A(\phi)$ is a bounded operator on $\mathcal{K}$ such that

$$\|A(\phi)\|_{\mathcal{K}} \leq \|\dot{\phi}\|_{L^\infty(\mathbb{R}^1)} + \|\phi\|_{L^\infty(\mathbb{R}^1)}. \tag{2.18}$$

(ii) For every $\phi_1 \in AC(\mathbb{R}^1)$ and $\phi_2 \in AC(\mathbb{R}^1)$ we have

$$A(\phi_1) + A(\phi_2) = A(\phi_1 + \phi_2) \tag{2.19}$$

and

$$A(\phi_1)A(\phi_2) = A(\phi_1\phi_2) = A(\phi_2)A(\phi_1). \tag{2.20}$$

(iii) If the sequence $\{\phi_n\}_{n=1}$ of elements of $AC(\mathbb{R}^1)$ pointwise converges to $\phi \in AC(\mathbb{R}^1)$, then

$$\underset{n\to+\infty}{\text{s-lim}}\ A(\phi_n) = A(\phi). \tag{2.21}$$

The convergence is understood as a strong convergence on $\mathcal{K}$ regarded as a Hilbert space.
(iv) For every $\phi \in AC(\mathbb{R}^1)$ the relation

$$A(\overline{\phi}) = A^{[*]}(\phi) \tag{2.22}$$

holds, where $\overline{\phi}$ is the complex conjugated function to ϕ.

We remark that

$$A(1) = I, \tag{2.23}$$

$1 \in AC(\mathbb{R}^1)$, holds. By equipping the set $AC(\mathbb{R}^1)$ with the norm $\|.\|_{AC(\mathbb{R}^1)}$,

$$\|\phi\|_{AC(\mathbb{R}^1)} = \|\phi\|_{L^\infty(\mathbb{R}^1)} + \|\dot{\phi}\|_{L^\infty(\mathbb{R}^1)}, \tag{2.24}$$

$\phi \in AC(\mathbb{R}^1)$, $AC(\mathbb{R}^1)$ forms a symmetric commutative Banach algebra with identity. The involution is given by $\phi \mapsto \overline{\phi}$.

According to M.A.Najmark and R.S.Ismagilov [37] the map $\phi \mapsto A(\phi)$ is a symmetric representation of the symmetric commutative Banach algebra $AC(\mathbb{R}^1)$ on the Krein space $\mathcal{K}$. As we can define such a symmetric representation of $AC(\mathbb{R}^1)$ on $\mathcal{R}_{-i} \times \mathcal{R}_i$ for every symmetric operator K satisfying (1.14) - (1.15), we call the map $\phi \to A(\phi)$ the associated representation of K.

It is unclear how characteristic is the associated representation. This means: Let $\phi \to A(\phi)$ and $\phi \to A'(\phi)$ be the associated representations of K and K' such that they are (J,J')-equivalent. If K and K' are completely non-selfadjoint, does it follow from this assumption that K and K' are unitarily equivalent?

Furthermore, the problem is unsolved if every symmetric representation $\phi \longrightarrow A(\phi)$ of the symmetric commutative Banach algebra $AC(\mathbb{R}^1)$ can be regarded as the associated representation of some symmetric operator K satisfying (1.14) and (1.15).

3. The associated group

We set $\phi_\sigma(t) = \exp(-i\sigma t)$, $\sigma, t \in \mathbb{R}^1$. The functions ϕ_σ belong to $AC(\mathbb{R}^1)$ for every $\sigma \in \mathbb{R}^1$. Hence, we can define

$$U(\sigma) = A(\phi_\sigma). \tag{3.1}$$

Theorem 3.1. The map $\mathbb{R}^1 \ni \sigma \mapsto U(\sigma) \in \mathcal{L}(\mathcal{K})$ forms a one-parameter strongly continuous group on $\mathcal{K}$ such that

$$\|U(\sigma)\|_{\mathcal{K}} \leq 1 + |\sigma| , \tag{3.2}$$

$$U^{[*]}(\sigma) = U(-\sigma), \tag{3.3}$$

$\sigma \in \mathbb{R}^1$.

Theorem 3.1 is a consequence of Theorem 2.1. A linear bounded operator U on a Krein space satisfying $U^{[*]} = U^{-1}$ is called a J-unitary operator. Consequently, $U(\sigma)$, $\sigma \in \mathbb{R}^1$, is a one-parameter strongly continuous group of J-unitary operators on $\mathcal{K}$. We call $U(\sigma)$, $\sigma \in \mathbb{R}^1$, the associated group of the symmetric operator K. The one-parameter strongly continuous group $U(\sigma)$, $\sigma \in \mathbb{R}^1$, has the representation

$$U(\sigma) = e^{-i\sigma L}, \tag{3.4}$$

$\sigma \in \mathbb{R}^1$, where L is called the associated generator of the symmetric operator K. The operator L is J-selfadjoint, i.e. $L^{[*]} = L$.

Corollary 3.2. The spectrum spec(L) of the associated generator L of the symmetric operator K is contained in $\mathbb{R}^1$, i.e. $\operatorname{spec}(L) \subseteq \mathbb{R}^1$.

This property is a consequence of (3.2).

4. Symmetric extensions

Now we want to find conditions in terms of the associated representation or the associated group which guarantee the existence of symmetric extensions $\hat{K}$ of K obeying (1.14) and (1.15). Naturally, we use the theory of J.von Neumannn on symmetric extensions of symmetric operators. In accordance with J.von Neumann, every symmetric extension $\hat{K}$ of a symmetric operator K is characterized by a partially isometric operator $\hat{V}$ acting from $F \subseteq \mathcal{R}_{-i}$ onto $G \subseteq \mathcal{R}_{i}$. The domain dom(K) is given by

$$\mathrm{dom}(K) = \mathrm{dom}(K) \dot{+} (\hat{V} + I)F, \tag{4.1}$$

i.e. every element $h \in \mathrm{dom}(\hat{K})$ has the unique representation

$$h = f + g_1 + \hat{V}g_1, \tag{4.2}$$

$f \in \mathrm{dom}(K)$, $g_1 \in F \subseteq \mathcal{R}_{-i}$. We introduce the projection operator

$$Q = \frac{1}{2}\begin{pmatrix} \hat{V}^*\hat{V} & \hat{V}^* \\ \hat{V} & \hat{V}\hat{V}^* \end{pmatrix} \tag{4.3}$$

on the Hilbert space $\mathcal{K}$, i.e. $Q^* = Q = Q^2$.

<u>Lemma 4.1.</u> Let $\hat{K}$ be a symmetric extension of K domain of which is given by (4.1). The symmetric extension $\hat{K}$ satisfies the conditions

$$M(\phi)\mathrm{dom}(\hat{K}) \subseteq \mathrm{dom}(\hat{K}) \tag{4.4}$$

and

$$\hat{K}M(\phi)f - M(\phi)\hat{K}f = -iM(\dot{\phi})f, \tag{4.5}$$

$f \in \mathrm{dom}(\hat{K})$, for every $\phi \in AC(\mathbb{R}^1)$ if and only if the associated representation of K obeys

$$A(\phi)Q = QA(\phi)Q \tag{4.6}$$

for every $\phi \in AC(\mathbb{R}^1)$.

Condition (4.6) means that $\mathcal{N} = Q\mathcal{K}$ is an invariant subspace of $A(\phi)$, $\phi \in AC(\mathbb{R}^1)$.

Now the notion of a neutral subspace of a Krein space will be crucial in the following. An element f of the Krein space $\mathcal{K}$ is called a neutral one if

$$[f,f] = 0. \tag{4.7}$$

A subspace $\mathcal{N}$ of $\mathcal{K}$ is called a neutral one if $\mathcal{N}$ consists of neutral elements. It is not hard to show that we have

$$[f,g] = 0 \tag{4.8}$$

for f and g arbitrary elements of a neutral subspace.

A neutral subspace $\mathcal{N}$ is said to be maximal neutral if $\mathcal{N}$ is not contained in any other neutral subspace of $\mathcal{K}$ except $\mathcal{N}$. Every maximal neutral subspace is either maximal non-negative or maximal non-positive or both.

If the last case happens , it means, if the neutral subspace is maximal non-negative and non-positive, then the subspace is called hypermaximal.

Now a simple calculation shows that $\mathcal{N} = Q\mathcal{K}$ is a neutral subspace. Hence we have established one direction of the following

<u>Theorem 4.2.</u> The symmetric operator K obeying (1.14) and (1.15) has a symmetric extension $\hat{K}$ obeying (4.4) and (4.5) if and only if the associated representation $A(\phi)$, $\phi \in AC(\mathbb{R}^1)$, of K has a non-zero neutral invariant subspace.

The converse is based on the following

<u>Lemma 4.3.</u> The orthogonal projection Q on $\mathcal{K}$ fulfils

$$Q^{[*]}Q = QQ^{[*]} = 0 \tag{4.9}$$

if and only if there is a partial isometry $\hat{V}$: $\mathcal{R}_{-i} \longrightarrow \mathcal{R}_i$ such that

$$Q = \frac{1}{2}\begin{pmatrix} \hat{V}^*\hat{V} & \hat{V}^* \\ \hat{V} & \hat{V}\hat{V}^* \end{pmatrix}. \tag{4.10}$$

The partial isometry is uniquely determined.

Theorem 4.2 admits a further refinement.

<u>Corllary 4.4.</u> The symmetric operator K has a maximal symmetric extension $\hat{K}$ obeying (4.4) and (4.5) if and only if the associated representation $A(\phi)$, $\phi \in AC(\mathbb{R}^1)$, of K has a maximal neutral invariant subspace. Furthermore, K has a selfadjoint extension obeying (4.4) and (4.5) if and only if $A(\phi)$, $\phi \in AC(\mathbb{R}^1)$, possesses a hypermaximal neutral invariant subspace.

We note that the results can be carried over the associated group.

<u>Corollary 4.5.</u> The symmetric operator K has a symmetric extension $\hat{K}$ obeying (4.4) and (4.5) if and only if the associated group $U(\sigma)$, $\sigma \in \mathbb{R}^1$, of K has a non-zero neutral invariant subspace. There is a maximal symmetric $\hat{K}$ if $U(\sigma)$, $\sigma \in \mathbb{R}^1$, has a maximal neutral invariant subspace and there is a selfadjoint K if and only if $U(\sigma)$ has a hypermaximal neutral invariant subspace.

The advantage of Corollary 4.5 is that the investigations can be reduced to a single operator, namely, the associated generator L.

5. Reduction

Let $\hat{K}$ be an extension of K such that the conditions (4.4) and (4.5) are fulfilled. If $\hat{K}$ is not selfadjoint, then we can associate with K a new associated representation $\hat{A}(\phi)$, $\phi \in AC(\mathbb{R}^1)$. Naturally, the question arises on the connection of the two associated representations $\hat{A}(\phi)$ and $A(\phi)$ of $\hat{K}$ and K, respectively. Let $\hat{K}$ be characterized by the partial isometry $\hat{V}: \mathcal{R}_{-i} \longrightarrow \mathcal{R}_i$. With $\hat{V}$ we connect the orthogonal projection P,

$$P = \begin{pmatrix} \hat{V}^*\hat{V} & 0 \\ 0 & \hat{V}\hat{V}^* \end{pmatrix}, \tag{5.1}$$

whose range coincides with F x G.

Proposition 5.1. Suppose that the symmetric extension $\hat{K} \supset K$ obeying (4.4) and (4.5) is given by the partial isometry $\hat{V}: \mathcal{R}_{-i} \longrightarrow \mathcal{R}_i$. If K is not selfadjoint, then the associated representation $\hat{A}(\phi)$ of $\hat{K}$ is defined on the Krein space $\hat{\mathcal{K}}$,

$$\hat{\mathcal{K}} = (I_{\mathcal{K}} - P)\mathcal{K}, \tag{5.2}$$

and can be obtained from the associated representation $A(\phi)$ of K by

$$\hat{A}(\phi) = (I_{\mathcal{K}} - P)A(\phi)\restriction(I_{\mathcal{K}} - P)\mathcal{K}, \tag{5.3}$$

$\phi \in AC(\mathbb{R}^1)$.

6. Concluding Remarks

The problem allows a further treatment assuming that the defect index of K is finite or one of the defect numbers $\dim(\mathcal{R}_{-i})$ and $\dim(\mathcal{R}_i)$ is finite. In these cases methods of the Pontragin space can be applied which allow far going conclusions. For instance it is possible to show that the closed symmetric operator K is selfadjoint if its defect index is finite ($\dim(\mathcal{R}_{-i}) < +\infty$, $\dim(\mathcal{R}_i) < +\infty$) and the conditions (1.14) - (1.16), (1.18) are fulfilled. Thus, if K obeys (1.14) - (1.16) and has a finite defect index, then the condition (1.18) must be violated at some points. It can be proved that these points coincide with the spectrum of the associated generator L. Furthermore, the problem can be deeper analyzed if the dimension of the Hilbert space $\mathfrak{h}$ is finite. But we must remark that our efforts end in failure assuming $\dim(\mathfrak{h}) = +\infty$ and an arbitrary defect index. Moreover, we have found that the results fail in this general case. For instance it is not true that the defect index of K arises only from a violation of (1.18). In the case $\dim(\mathfrak{h}) = +\infty$ it can happen that all the conditions (1.14) - (1.16), (1.18) are fulfilled but K is not selfadjoint. Unfortunately, the last case is important in application to evolution equations.

References:

[1] T.Kato:Integration of the equation of evolution in Banach space, J. Math. Soc. Japan 5(1953), 208 - 234.

[2] T.Kato: On linear differential equations in Banach spaces, Comm. Pure Appl. Math. 9(1956) 479 - 486.

[3] T.Kato: Abstract evolution equation of parabolic type in Banach and Hilbert spaces, Nagoya Math. J. 19(1961),93 - 125.

[4] T.Kato, H.Tanabe: On the abstract evolution equation, Osaka Math. J. 14(1962), 107 - 133.

[5] T.Kato: Linear evolution equations of "hyperbolic" type, J. Fac. Science, University of Tokyo, Sect. IA, Mathematics, 17(1970), 241 - 258.

[6] T.Kato: Linear evolution equations of "hyperbolic" type, II, J. Math. Soc. Japan 25(1973), 648 - 666.

[7] M.Reed, B.Simon:"Methods of Modern Mathematical Physics I: Fourier Analysis, Selfadjointness", Academic Press, New-York-San Francisco-London 1975.

[8] K.Yajima: Existence of solutions for Schrödinger evolution equations, Comm. Math. Physics 110(1987), 415 - 426.

[9] H.Neidhardt: "Integration von Evolutionsgleichungen mit Hilfe von Evolutionshalbgruppen", Dissertation, AdW der DDR, Berlin 1979.

[10] H.Neidhardt: On abstract linear evolution equations, II, Preprint, AdW der DDR, Institut für Mathematik, P-MATH-07/81, Berlin 1981.

[11] O.A.Ladyženskaja: On the solution of operator equations of different types, Doklady Akad. Nauk SSSR 102(1955), 207 - 210 (in Russian).

[12] O.A.Ladyženskaja: On the solution of non-stationary operator equations, Mat. Sbornik 39(1956), 491 - 524(in Russian).

[13] J.L.Lions: Equations différentielles à coefficients opèrateurs non bornés, Bull. Soc. Math. France 86(1958), 321 - 330.

[14] J.L.Lions: Sur certaines équations aux derivées partielles à coefficients opèrateurs non bornés, J. Anal. Math. 6(1958), 333 - 355.

[15] J.L.Lions: Equations différentielles du premier ordre dans un espace de Hilbert, C.R. Acad. Sci. Paris, série A, 248(1959), 1099 - 1102.

[16] G.DaPrato: Weak solutions for linear abstract differential equations in Banach spaces, Advances in Mathematics 5(1970), 181 - 245.

[17] G.DaPrato: Sums of linear operators, in: "Linear Operators and Approximation, II", Proceedings of the Conference held at the Oberwolfach Mathematical Research Institute, Black Forest, March 30 - April 6, 1974, pp. 461 - 472 (eds. P.L.Butzer, B.Sz.-Nagy).

[18] M.Iannelli: On the Green function for abstract evolution equations, Bolletino U.M.I. (4) 6(1972), 154 - 174.

[19] G.DaPrato, P.Grisvard: Sommes d'opérateurs linéaires et équations différentielles opérationelles, J. Math. Pures Appliquées 54(1975), 305 - 387.

[20] L.Paquet: Equations d'evolution pour opérateurs locaux et équations aux derivées partielles, C.R. Acad. Sci. Paris, serie A, 284(1977).

[21] J.S.Howland: Stationary scattering theory for time-dependent hamiltonians, Math. Ann. 207(1974), 315 - 335.

[22] H.Neidhardt: On abstract linear evolution equations, I, Math. Nachr. 103(1981), 283 - 298.

[23] H.Neidhardt: On abstract linear evolution equations, III, Preprint, AdW der DDR, Institut für Mathematik, P-MATH-05/82, Berlin 1982.

[24] M.Reed, B.Simon:"Methods of Modern Mathematical Physics I: Functional Analysis", Academic Press, New-York-San Francisco-London 1974.

[25] B.Fuglede: On the relation PQ - QP = -iI, Math. Scand. 20(1967), 79 - 88.

[26] N.S.Poulsen: On the canonical commutation relations, Math. Scand. 32(1973), 112 - 122.

[27] P.E.T.Jørgensen: Selfadjoint operator extensions satisfying the Weyl commutation relations, Bull. (New Series) Amer. Math. Soc. 1(1979), 1, 266 - 269.

[28] P.E.T.Jørgensen, P.S.Muhly: Selfadjoint extensions satisfying the Weyl operator commutation relations, J. d'Analyse Math. 37(1980), 46 - 99.

[29] W.J.Phillips: On the relation PQ - QP = -iI, Pac. J. Math. 95(1981), 435 - 441.

[30] K.Schmüdgen: On the Heisenberg commutation relation I, J. Funct. Analysis 50(1983), 8 - 49.

[31] K.Schmüdgen: On the Heisenberg commutation relation II, Publ. RIMS, Kyoto Univ. 19(1983), 601 - 671.

[32] G.Dorfmeister, J.Dorfmeister: Classification of certain pairs of operators (P,Q) satisfying [P,Q] = -iId, J. Funct. Analysis 57(1984), 301 - 328.

[33] H.Neidhardt: Symmetric extensions preserving additional conditions, Preprint, AdW der DDR, Institut für Mathematik, P-MATH-16/82, Berlin 1982.

[34] N.I.Achieser, I,M.Glasmann: "Theorie der linearen Operatoren im Hilbert-Raum", Akademie-Verlag, Berlin 1975.

[35] J.Bognar: "Indefinite inner product spaces", Ergebnisse der Mathematik und ihrer Grenzgebiete, Bd. 78, Springer Verlag, Berlin-Heidelberg-New York 1974.

[36] T.Ja.Azizov, I.S.Jodvidov: "Basics facts about linear operators in spaces with indefinite metric", Izd. "Nauka", Moskva 1986 (in Russian).

[37] M.A.Naimark, R.S.Ismagilov: Representation of groups and algebras with indefinite metric, Itogi nauki, mat. anal. 1968, AN SSSR, Institute Sci. Information, Moskva 1969.

ENERGY-DEPENDENT INTERACTIONS AND THE EXTENSION THEORY

K.A.Makarov
Physical Institute, Leningrad State University,
1 Maya 100, Petrodvoretz Leningrad 198904

The aim of this lecture is to present a short summary of some aspects a mathematical construction developed at Leningrad University for two-body exactly solvable models with point interactions [1-3, 5-9].

1. M.G.Krein's formula for the generalized resolvent and boundary-form representation

We start in this paper with a reformulation of von Neumann's theory of self-adjoint extensions in terms of boundary forms which based on the following theorem [2]:

Theorem 1 Let $\mathfrak{N}_i$ be a deficiency subspace corresponding to $\lambda = i$ of a densely defined symmetric operator A_0 in a Hilbert space H with equal deficiency indices. Let A be a self-adjoint extension of A_0. Then the adjoint operator A_0^* can be described as

$$\mathrm{Dom}(A_0^*) = \{ u: u = (A-iI)^{-1}\vartheta + A(A-iI)^{-1}\mathcal{E}^+ + \tag{1}$$

$$+ (A-iI)^{-1}\mathcal{E}^-, \quad \mathcal{E}^\pm \in \mathfrak{N}_i, \quad \vartheta \in \mathfrak{N}_i^\perp \}.$$

Hence to each $u \in \mathrm{Dom}(A_0^*)$, vectors $\mathcal{E}^\pm(u)$ are associated.

The boundary form of the adjoint operator A_0^* can be written in terms of these "boundary values" as

$$\langle A_0^* u, v\rangle - \langle u, A_0^* v\rangle = \langle \mathcal{E}^-(u), \mathcal{E}^+(v)\rangle -$$

$$- \langle \mathcal{E}^+(u), \mathcal{E}^-(v)\rangle \equiv J(\mathcal{E}(u), \mathcal{E}(v)). \tag{2}$$

The self-adjoint extensions of A_0 are in one-to-one correspondance with the bundle of Lagrange planes of the boundary form, i.e., the planes in $\mathfrak{N}_i + \mathfrak{N}_{-i}$ where the boundary form (2) vanishes. All such Lagrange planes can be reduced to the surface in $\mathfrak{N}_i + \mathfrak{N}_{-i}$, defined as follows

$$L_\Gamma = \left\{ u \in \mathrm{Dom}(A_0^*): \quad \mathcal{E}^-(u) = \Gamma \mathcal{E}^+(u) \right\} \tag{3}$$

where Γ is a self-adjoint operator in $\mathfrak{N}_i$. Thus

$$J(\mathcal{E}(u), \mathcal{E}(v)) = 0$$

for any $u, v \in L_\Gamma$. Finally the self-adjoint extension A_Γ of the operator A_0 is determined as

$$A_\Gamma = A_0^* \big|_{L_\Gamma} \tag{4}$$

Let A_0 be a restriction of a self-adjoint operator A in H. Then for operators A and A_Γ the following theorem (see [2]) holds

<u>Theorem 2</u> The resolvents of self-adjoint extensions A, A_Γ of the symmetric operator A_0 are related by

$$(A_\Gamma - \lambda I)^{-1} = (A - \lambda I)^{-1} + \frac{A+iI}{A-iI} P \left(\Gamma - P \frac{I+\lambda A}{A-\lambda I} P\right)^{-1} P \times \frac{A-iI}{A-\lambda I} \tag{5}$$

where P denotes the projection onto $\mathfrak{N}_i$ in H.

The relation (5) is known as M.G.Krein's formula [4]. It makes many interesting model problems in quantum mechanics solvable [2,3,6-9].

Now we are going to apply the above results to the two-channel situation: $H = H^{ex} \oplus H^{in}$, $A = A^{ex} \oplus A^{in}$, $\mathfrak{N}_i = \mathfrak{N}_i^{ex} \oplus \mathfrak{N}_i^{in}$. Let us denote

$$D^{ex,in}(\lambda) = P \frac{I+\lambda A}{A-\lambda I} P \big|_{H^{ex,in}} \tag{6}$$

Then we can define the operator $D(\lambda)$ in H as

$$D(\lambda) = \begin{pmatrix} D^{ex}(\lambda) & 0 \\ 0 & D^{in}(\lambda) \end{pmatrix} \tag{7}$$

One can notice that $D(\lambda)$ is an operator-valued analytical function with positive imaginary part in the upper halfplane (the so-called R-function). It is the main functional parameter of the method discussed here.

<u>Lemma 1</u> The "boundary values" $\mathcal{E}^{\pm}(u)$ of the solution to the adjoint equation $(A_0^* - \lambda I)u = f$ are related by

$$\mathcal{E}^{-}(u) = P\,\frac{A-iI}{A-\lambda I}\,f + D(\lambda)\,\mathcal{E}^{+}(u) \tag{8}$$

where $\mathcal{E}^{\pm}(u) = \mathcal{E}^{\pm}_{ex}(u) \oplus \mathcal{E}^{\pm}_{in}(u)$.

In the two-channel case, $H = H^{ex} \oplus H^{in}$, the parametrizing operator Γ has a block structure:

$$\Gamma = \begin{pmatrix} \Gamma_{ee} & \Gamma_{ei} \\ \Gamma_{ie} & \Gamma_{ii} \end{pmatrix}, \quad \Gamma_{ei} = \Gamma^{+}_{ie}, \quad \Gamma_{ee} = \Gamma^{+}_{ee}, \quad \Gamma_{ii} = \Gamma^{+}_{ii}.$$

The resolvent $R_\Gamma(\lambda) = (A_\Gamma - \lambda I)^{-1}$ has a block structure too; its components can be reconstructed from the "pure external" block $R^{ee}_\Gamma(\lambda)$ alone [8]. Lemma 1 allows to obtain for $R^{ee}_\Gamma(\lambda)$ the following formula

$$R^{ee}_\Gamma(\lambda) = (A^{ex} - \lambda I^{ex})^{-1} + \frac{A^{ex}+iI^{ex}}{A^{ex}-\lambda I^{ex}}\, Q^{ex}(\lambda)\, \frac{A^{ex}-iI^{ex}}{A^{ex}-\lambda I^{ex}} \tag{9}$$

where I^{ex} is the identity operator in H^{ex} and $Q^{ex}(\lambda)$ is the operator-valued R-function in $\mathfrak{N}_i$:

$$Q^{ex}(\lambda) = \left\{(\Gamma_{ee} - D^{ex}(\lambda)) - \Gamma_{ei}(\Gamma_{ii} - D^{in}(\lambda))^{-1}\Gamma_{ie}\right\}^{-1}. \tag{10}$$

The formulae for resolvents of self-adjoint extensions are valid also for non-densely defined symmetric operator A_0 with equal deficiency indices, in particular for finite-dimensional internal operators, $\dim H^{in} < \infty$ (see the contribution of B.S.Pavlov to this volume).

2. Zero-range potential models and some generalizations

The simplest example of exactly solvable quantum model based on Krein's formula is represented by point interaction of particles with an internal structure [2,3,6-9].

Let A^{in} be some self-adjoint operator in Hilbert space H^{in} and let θ be a generating element of A^{in}:

$$H^{in} = \overline{\bigvee_t \exp(iA^{in}t)\ \theta} \tag{11}$$

The restriction of A^{in} to the domain

$$\mathrm{Dom}(A_0^{in}) = (A^{in} - iI)^{-1}\{\theta\}^{\perp}$$

will be denoted as A_0^{in}. The only deficiency vector of the operator A_0^{in} corresponding to the point $\lambda = i$ coincides with θ . If the domain $\mathrm{Dom}(A_0^{in})$ is dense in H^{in}, i.e., if $\theta \notin \mathrm{Dom}(A)$, then the domain $\mathrm{Dom}(A_0^{in\,*})$ of the adjoint operator $A_0^{in\,*}$ is given by the formula (1), where $\mathcal{N}_i = \{\theta\}$, $\varepsilon^{\pm} = \zeta^{\pm}\theta$ and $\zeta^{\pm} \in \mathbf{C}$. The boundary form (2) now looks like

$$J^{in}(u, v) = \zeta^{-}(u)\overline{\zeta^{+}(v)} - \zeta^{+}(u)\overline{\zeta^{-}(v)} . \tag{12}$$

Let A^{ex} be Laplace operator in $\mathbb{R}^3$ and $A_0^{ex} = (-\Delta)_0$ is its restriction to the set of all smooth functions u, $u \in L_2(R^3)$, vanishing in the neighbourhood of the origin. Then the domain of the adjoint operator is [3]

$$\mathrm{Dom}(A_0^{ex\,*}) = \left\{u \in W_2^2(R^3/\{0\}):\ u(x) = \frac{u_0}{4\pi|x|} + u_1 + o(1)\right\} \tag{13}$$

and the external boundary form can be written in the terms of the asymptotic boundary values u_0, u_1:

$$J^{ex}(u, v) = -u_0\overline{v_1} + u_1\overline{v_0} . \tag{14}$$

The set of the self-adjoint extensions of the operator $(-\Delta_0) \oplus \oplus A_0^{in}$ is parametrized by Lagrange planes of the combined boundary form $J = J^{ex} + J^{in}$. The simplest realization of such a Lagrange plane is given by the boundary condition

$$\begin{pmatrix} u_0 \\ \zeta^- \end{pmatrix} = \Gamma \begin{pmatrix} u_1 \\ \zeta^+ \end{pmatrix} \tag{15}$$

with some Hermitian 2x2-matrix Γ.

Let us calculate the external components of scattered waves which are the eigenfunctions of the absolutely-continuous spectrum of the self-adjoint operator corresponding to the boundary condition (15). These components look as follows

$$u_e = e^{-ik\langle x,\vartheta\rangle} + f(k)\frac{e^{ik|x|}}{4\pi|x|}, \quad k = \sqrt{\lambda} > 0 \tag{16}$$

The boundary condition (15) and Lemma 1 lead to the relation

$$\begin{pmatrix} 1+\frac{ikf(k)}{4\pi} \\ D^{in}(\lambda)\zeta^+ \end{pmatrix} = \Gamma \begin{pmatrix} f(k) \\ \zeta^+ \end{pmatrix}, \quad \Gamma = \begin{pmatrix} \Gamma_{00} & \Gamma_{01} \\ \Gamma_{01} & \Gamma_{11} \end{pmatrix}, \quad \Gamma_{jj} \in \mathbb{R}. \tag{17}$$

Hence we get the following expressions for scattering amplitude f(k) and s-wave scattering matrix $S^0(k)$:

$$f(k) = \left\{ \left(\Gamma_{00} - \frac{ik}{4\pi}\right) - |\Gamma_{01}|^2 (\Gamma_{11} - D^{in}(k^2))^{-1} \right\}^{-1}, \tag{18}$$

$$S^0(k) = \frac{(\Gamma_{00} + ik/4\pi) - |\Gamma_{01}|^2 (\Gamma_{11} - D^{in}(k^2))^{-1}}{(\Gamma_{00} - ik/4\pi) - |\Gamma_{01}|^2 (\Gamma_{11} - D^{in}(k^2))^{-1}}. \tag{19}$$

The scattering matrix obtained above is non-trivial in the s-channel only. Our approach fails for higher momenta l = 1,2,..., because the corresponding multipols are not square integrable. There exist two ways of constructing of solvable models of an atom with higher momenta. The first one is based on use of an indefinite metric [10] . The other can be done in $L_2(R^3)$ but one has to sacrify locality of the interaction [5,8,9].

Let $(-\Delta)_0$ be the restriction of $(-\Delta)$ to the linear set of all smooth functions vanishing on the sphere $\Sigma_R = \{x: |x| = R\}$. The domain of the adjoint operator $(-\Delta)_0^*$ consists of functions u(x):

$$u(x) = u_0(x) + \frac{1}{4\pi}\int |x-s|^{-1} \exp(ik|x-s|)\,\rho_u(s)\,ds \tag{20}$$

where $u_0 \in Dom(-\Delta)_0$ and $\rho_u \in W_2^{-3/2}(\Sigma_R)$. The boundary form is equal to

$$J(u, v) = \iint_{\Sigma_R} \left\{(-u)\overline{\frac{\partial v}{\partial n}} - \frac{\partial u}{\partial n}\overline{(-v)}\right\} ds = \tag{21}$$
$$= \int_{\Sigma_R} \left\{(-u)\overline{\rho_v} - \rho_u\overline{(-v)}\right\} ds.$$

Let A^{in} be a Hermitian operator in an internal Hilbert space H^{in}. We denote by A_0^{in} some symmetric restriction of A^{in} with the deficiency subspace $\mathrm{def}_+(A_0^{in}) = \mathfrak{N}_i^{in}$ corresponding to the point $\lambda = i$, $\dim \mathfrak{N}_i^{in} = n$. Then the sought self-adjoint extension of the operator $(-\Delta)_0 \oplus A_0^{in}$ can be constructed as Friedrichs extension of the below bounded operator A_Γ. The operator A_Γ is defined by the boundary conditions

$$\begin{pmatrix} -[\partial_n u]\big|_{\Sigma_R} \\ \mathcal{E}^+ \end{pmatrix} = \begin{pmatrix} \Gamma_{00} & \Gamma_{01} \\ \Gamma_{10} & \Gamma_{11} \end{pmatrix} \begin{pmatrix} u\big|_{\Sigma_R} \\ \mathcal{E}^- \end{pmatrix} \tag{22}$$

Here $\mathcal{E}^\pm \in \mathfrak{N}_i^{in}$, $\Gamma_{ik} = \Gamma_{ki}^*$. We obtain the simplest picture of p-scattering by taking $\Gamma_{01} = \gamma Y_{1m}$, $\Gamma_{10} = \bar{\gamma}\langle \cdot, Y_{1m}\rangle$, $\Gamma_{00} = \Gamma_{11} = 0$, Y_{1m} is one of the l=1 spherical functions on Σ_1 and $\gamma \in C^n$. This leads to the boundary condition

$$[\partial_n u]\big|_{\Sigma_R} = -(\gamma, D^{in}(\lambda)\bar{\gamma})\langle u|_{\Sigma_R}, Y_{1m}\rangle Y_{1m} \tag{23}$$

where $(\cdot, \cdot)$ means the scalar product in $C^n \sim \mathfrak{N}_i^{in}$ and $\langle \cdot, \cdot \rangle$ means the scalar product in $L_2(\Sigma_R)$.

The boundary condition (23) can be reformulated in terms of the generalized energy-dependent potential $V(\lambda)$ [8]:

$$V(\lambda): f_0 \mapsto -(\gamma, D^{in}(\lambda)\bar{\gamma})\langle f_0|_{\Sigma_R}, Y_{1m}\rangle Y_{1m}\delta_{\Sigma_R}, \tag{24}$$
$$f_0 \in L_2(R^3).$$

Here δ_{Σ_R} means δ-function with support at the sphere Σ_R. The form of the potential $V(\lambda)$ allows us to obtain an explicit solution for the Lippmann-Schwinger equation for the resolvent

$$G(\lambda) = G_0(\lambda) - G_0(\lambda)\, V(\lambda)\, G(\lambda) \tag{25}$$

This solution looks like

$$G(\lambda) = G_0(\lambda) + \langle G_0(\lambda)\cdot, Y_{1m}\rangle G_0(\lambda)\, Y_{1m} \times$$
$$\times \left\{ (\gamma, D^{in}(\lambda)\bar{\gamma})^{-1} - \langle G_0(\lambda) Y_{1m}, Y_{1m}\rangle \right\}^{-1} \tag{26}$$

We denote here $G_0(\lambda)$ the free resolvent of the Laplace operator in R^3:

$$G_0(\lambda) = (-\Delta - \lambda)^{-1} .$$

Now we can obtain the expression for the external component of the scattered wave as the coefficient of the spherical wave in the asymptotics of the Green function $G(x,x',\lambda)$ when $|x'| \to \infty$ [11]. The free Green function looks like

$$G_0(x,x',\lambda) = \frac{i}{4} \frac{\lambda^{1/4}}{\sqrt{2\pi}} \frac{H^{(1)}_{1/2}(\sqrt{\lambda}\,|x-x'|)}{|x-x'|^{1/2}}$$

and has the asymptotics

$$G_0(x,x',\lambda) \underset{|x'|\to\infty}{\sim} e^{-i\langle x,\hat{x}'\rangle\sqrt{\lambda}} \frac{e^{i\sqrt{\lambda}\,|x'|}}{4\pi\,|x-x'|} .$$

This gives the asymptotical expression for the $G(x,x',\lambda)$, defined by (26):

$$G(x,x',\lambda) \underset{|x'|\to\infty}{\sim} \frac{e^{i\sqrt{z}\,|x'|}}{4\pi\,|x-x'|}\, u^0(x,\vec{k}) \quad ; \; k = |\vec{k}| = \sqrt{\lambda} \; ;$$
$$\hat{k} = \frac{\vec{k}}{k} = -\hat{x}' \tag{27}$$

where $u^0(x,\vec{k})$ is the external component of the scattered wave with a momentum $\vec{k}$:

$$u^0(x,\vec{k}) = e^{i\langle x,k\rangle} + \int_{\Sigma_R} e^{i\sqrt{\lambda}R\langle\hat{s},\hat{k}\rangle}\, Y_{1m}(\hat{s})\, d\hat{s} \;\times$$
$$\times \frac{G_0(x,R\hat{s},k^2) Y_{1m}(\hat{s})\, d\hat{s}}{(\gamma, D^{in}(k^2)\bar{\gamma})^{-1} - \int_{\Sigma_R}\int_{\Sigma_R} d\hat{s}\, d\hat{s}'\, G_0(R\hat{s}',R\hat{s},k^2) Y_{1m}(\hat{s})\overline{Y_{1m}(\hat{s}')}} \tag{28}$$

The asymptotics of $u^0(x,\vec{k})$ when $|x| \to \infty$,

$$u^0(x,\vec{k}) \underset{|x|\to\infty}{\sim} e^{i\langle x,k\rangle} + f(\hat{x},\vec{k})\, e^{ik|x|}/(4\pi|x|) ,$$

leads to the following scattering amplitude $f(\hat{x},\vec{k})$:

$$f(\hat{x},\vec{k}) = \int_{\Sigma_R} e^{ikR\langle\hat{s},\hat{k}\rangle} Y_{1m}(\hat{s})\, d\hat{s} \int_{\Sigma_R} e^{ikR\langle\hat{s},\hat{x}\rangle} \overline{Y_{1m}(\hat{s})}\, d\hat{s} \times$$
$$\times\left\{(\gamma, D^{in}(k^2)\bar{\gamma})^{-1} - \int_{\Sigma_R}\int_{\Sigma_R} d\hat{s}\, d\hat{s}'\, Y_{1m}(\hat{s})\overline{Y_{1m}(\hat{s}')}\, G_0(\hat{s}',\hat{s},k^2)\right\}^{-1} \quad (29)$$

Finally, for the S-matrix we get:

$$S(x,k) = 1 + \frac{ik}{2\pi} f(x,k) \quad (30)$$

where $f(x,k)$ is given by (29).

Thus we have demonstrated that one can use no indefinite metric even in the case $l \geq 1$, and obtained S-matrix (30) has rather complicate analytical structure. All the results were practically obtained on the algebraic level of computations.

3. Boundary conditions on thin manifolds

The next natural question is: how one can generalize the scheme of the boundary conditions model to the case of higher codimension ("thin") manifolds ? The answer can be obtained using the embedding theorems which allow to state boundary conditions on manifolds of lesser dimensions.

Let us consider the Laplacian $-\Delta$ in R^n and let L^m be a hypersurface in R^n, $\dim L^m = m$. Restrict $-\Delta$ to the symmetric operator $(-\Delta)_0$ specified by its domain

$$\mathrm{Dom}(-\Delta)_0 = C_0^\infty(R^n \setminus L^m) \quad (31)$$

Notice that $(-\Delta)_0$ is e.s.a. if codim L^m in $R^n \geq 4$. For codim $L^m < 4$, the deficiency subspace corresponding to $\lambda = i$ of $(-\Delta)_0$ consists of the elements

$$\theta_{\rho_p}(x) = \int_{L^m} (\partial n)^p\, G(x-s,i)\, \rho_p(s)\, ds \quad (32)$$

with such distribution $\rho_p(s)$ for wich $\theta_{\rho_p}(x) \in L_2(R^n)$. Here

$G(x,s,i)$ means the Green function of $(-\Delta)$, ∂n is the normal derivative on L^m. In order to find the class of ρ_p with $\theta_{\rho_p} \in L^2(R^n)$ or $\theta_{\rho_p} \in W_2^1(R^n)$, we use the momentum representation:

$$\rho_p(s) \rightarrow \hat{\rho}_p(k_\parallel) .$$

Simple calculations show that

$$\int \left| \frac{k_\perp^p}{k_\parallel^2 + k_\perp^2 - i} \hat{\rho}_p(k_\parallel) \right|^2 dk_\perp \, dk_\parallel = \tag{33}$$

$$= \int | \hat{\rho}_p(k_\parallel) |^2 dk_\parallel \int \frac{k_\perp^{2p}}{[k_\parallel^2 + k^2]^2 - i} dk_\perp$$

Here $k_\parallel$ denotes the momentum tangential to L^m and $k_\perp \perp k_\parallel$. The following assertion is valid

<u>Lemma 2</u> Suppose that

$$\int \frac{k_\perp^{2p}}{[k_\parallel^2 + k_\perp^2]^2 - i} dk_\perp < \infty \tag{34}$$

Then $\theta_{\rho_p} \in L_2(R^6)$ if codim $L^m < 4-2p$ and ρ_p satisfies the following conditions:

codim L^m	ρ_0 belongs to the class	ρ_1 belongs to the class
1	$W_2^{-3/2}$	$W_2^{-1/2}$
2	W_2^{-1}	not exist
3	$W_2^{-1/2}$	not exist

Here W_2^s are Sobolev classes of distributions. For $p \geqslant 2$ the corresponding Sobolev class does not exist.

If one is interested in the situation $\theta_{\rho_p} \in W_2^1(R^6)$, it is necessary to use the $p=0$ distribution only and $\rho_0 \in W_2^{-1/2}(R^5)$.

4. Three- body quantum problem with point interactions

It is interesting to apply the above technique to the three-body problem with point interactions. After the separation of the centre of masses the configurational space of the three-dimensional three-body system is $L_2(R^6)$. There are three 3-dimensional planes in R^6 corresponding to the pair point interactions. We denote L the union of them. Hence, $H = L_2(R^6)$, $\dim L = \operatorname{codim} L = 3$, $\mathcal{L} = -\Delta$, $\mathrm{Dom}(\mathcal{L}) = W_2^2(R^6)$.

Let us restrict the Hamiltonian $-\Delta$ to the symmetric operator $\mathcal{L}_0 = (-\Delta)_0$ with the domain

$$\mathrm{Dom}(-\Delta)_0 = C_0^\infty(R^6 \setminus L) .$$

According to Lemma 2 the deficiency subspace of $\mathcal{L}_0$ corresponding to $\lambda = i$ coincides with the linear set of functions

$$\mathcal{E} = \int_L G_i \rho_0 \, ds \quad , \quad \rho_0 \in W_2^{-1/2}(L) \quad . \tag{35}$$

The Friedrichs extension of $\mathcal{L}_0$ leads to the original operator $\mathcal{L}$. That is why we can use procedure described above. Namely, let us consider the linear set

$$\begin{aligned} \mathcal{V} = \Big\{ u = \tilde{u} + \frac{\mathcal{L}}{\mathcal{L} - i} \mathcal{E}^+ + (\mathcal{L} - i)^{-1} \mathcal{E}^- ; \\ \mathcal{E}^+ = (\mathcal{L} + i)^{-1} \rho_0^+ \quad , \\ \mathcal{E}^- = (\mathcal{L} + i)^{-1} \rho_0^- \quad , \\ \tilde{u} \in \mathrm{Dom}(-\Delta)_0 \Big\} \end{aligned} \tag{36}$$

and restrict the operator $(-\Delta)_0^*$ to $\mathcal{V}$. Then the following lemma is valid.

<u>Lemma 3</u> Let $u, v \in \mathcal{V}$. Then

$$\begin{aligned} \langle \mathcal{L}_0^* u, v \rangle - \langle u, \mathcal{L}_0^* v \rangle = \langle \mathcal{E}^-(u), \mathcal{E}^{+}(v) \rangle_{L_2(R^6)} - \\ \langle \mathcal{E}^+(u), \mathcal{E}^-(v) \rangle_{L_2(R^6)} = \langle \operatorname{Im} G_i \rho_0^-(u), \rho_0^+(v) \rangle_{L_2(L)} - \\ - \langle \rho_0^+(u), \operatorname{Im} G_i \rho_0^-(v) \rangle_{L_2(L)} \quad . \end{aligned} \tag{37}$$

So we have the natural variables (conjugated in the sense of Sobolev classes) of the symplectic form:

$$u^- \equiv \mathrm{Im}\, G_i \rho_0^-(u) \quad ,$$

$$\rho_0^+(u) \; .$$

How one can calculate ρ_0^+, u^- for a given $u \in \mathcal{V}$? The answer is given by

<u>Lemma 4</u> The integral $\int_\Omega \rho_0^+(u)\, dy$ for any Ω is given by

$$\int_\Omega \rho_0^+(u)\, dy = \lim_{\delta \to 0} \int_{\Omega_\delta} \partial n\, u\, ds$$

$$\Omega_\delta = \{ X \in R^6 : \mathrm{dist}(X, \Omega) = \delta \}, \tag{38}$$

and

$$u^- = \mathrm{Im}\, G_i \rho_0^- = \lim \, (u(X) - \int_L \mathrm{Re}\, G_i \rho_0^+(u)\, dy)$$

On the base of Lemma 4 we can write the boundary condition

$$\begin{pmatrix} u^- \\ \varepsilon^- \end{pmatrix} = \Gamma \begin{pmatrix} \rho^+ \\ \varepsilon^+ \end{pmatrix} \tag{39}$$

for the thin manifold L in R^6. Such constructions have been used in [5] for constructing the below bounded Hamiltonian in three-body system with point interactions and internal structure. As it was shown in [5] one can not construct such a self-adjoint operator by the extension theory methods without using an additional Hilbert space.

Thus in this situation rather easy methods based on von Neumann theory allow to solve rather hard and important problem related to the many-body systems with singular interactions.

Acknowledgements

I want to express my gratitude to B.S.Pavlov, Yu.A.Kuperin and Yu.B.Melnikov for the assistance in this work.

References

1. V.M.Adamjan, B.S.Pavlov, Zap.Nauch.Sem.LOMI 149 (1986), 7.
2. B.S.Pavlov, Uspekchi Matem.Nauk 42 (1987), 99.
3. B.S.Pavlov, Teor.Mat.Fiz. 59 (1984), 345.
4. M.G.Krein, Dokl.Akad.Nauk SSSR, 52 (1946), 657.
5. B.S.Pavlov, Matem.Sbornik, 1988 (to appear).
6. Yu.A.Kuperin, K.A.Makarov, B.S.Pavlov, Teor.Mat.Fiz. 69 (1986), 100.
7. Yu.A.Kuperin, K.A.Makarov, Yu.B.Melnikov, Teor.Mat.Fiz. 74 (1988), 103.
8. Yu.A.Kuperin, K.A.Makarov, S.P.Merkuriev et al., ITP-Budapest Report N441, Budapest, 1986.
9. Yu.A.Kuperin, K.A.Makarov, Yu.B.Melnikov: in "Theory of Quantum Systems with Strong Interactions", Kalinin, 1987, p.63.
10. Yu.G.Shondin, Teor.Mat.Fiz. 64 (1985), 432.
11. S.P.Merkuriev, L.D.Faddeev: Quantum Scattering Theory in Few-Body Systems, Moscow, Nauka, 1985 (in Russian).

ON PERTURBATIONS FOR SELF-ADJOINT GENERATORS OF FELLER PROCESSES

M. Demuth
Institute of Mathematics, Mohrenstr. 39, 1086 Berlin-Mitte, G.D.R.

1. Introduction

The aim of this article is to explain some spectral theoretical consequences for generators of Feller processes if they are perturbed by regular and singular potentials. The content is based on both the book by van Casteren [1] on generators of strongly continuous semigroups and on the report by Demuth [2] on scattering theory for generators of Markov processes. In both monographs Feller semigroups are introduced and investigated. In [1] the Feynman-Kac formalism for regular potentials is studied extensively. In [2] strongly continuous semigroups with singular potentials over unbounded regions are included. In the present article some further connections between these aspects in the Feller semigroup theory are sketched.

In section[2] the Feller semigroups are introduced in $L^{\infty}(\mathbb{R}^n)$. This definition is extended to $L^2(\mathbb{R}^n)$ such that the Feller generators become selfadjoint. Regular and singular perturbations are considered in sections 3 and 4, respectively.The Kato-class potentials are described. The singularity regions are classified. In section 5 sufficient conditions are given, such that certain differences of Feller semigroups are Hilbert-Schmidt or trace class operators. The conditions are satisfied even in the case of star-like singularity regions.

For the sake of shortness the theory is explained roughly, proofs are omitted, if they are known from the literature, and sketched for the new results. Moreover the content of the article is compressed. The interesting reader is referred to a more detailed version of this article which will be published in the Mathematische Nachrichten.

The results were presented at the workshop on "Applications of the Selfadjoint-Extensions Theory in Quantum Physics", Dubna, 1987. The author thanks Dr. P. Exner for the kind invitation and hospitality during this workshop.

2. Unperturbed generators of Feller processes

The main connection between the stochastic and spectral analysis is given by the theory of strongly continuous semigroups. Therefore, at first, it will be introduced the semigroup of the Feller process, i. e. the Feller semigroup, considered in this article.

Denotation 1: Probability space (see [3])

Let $(\Omega_x, \mathcal{B}_x, P_F)$ be a probability space, where Ω_x is the set of all continuous functions $\omega(\cdot)$, mapping $[0,\infty)$ into $\mathbb{R}^n$, starting at x. P_F is a probability measure on $\mathcal{B}_x$.

Assumption A: Let $(\Omega_x, \mathcal{B}_x, P_F, \omega(\cdot))$ be a Feller process homogeneous in time with the transition function

$$P(t,x,E) := P_F \{\omega : \omega(t) \in E , \omega(0) = x\} ,$$

$x \in \mathbb{R}^n$, $t \in [0,\infty)$, E a Borel set in $\mathbb{R}^n$. Assume a transition density function p, mapping $(0,\infty) \times \mathbb{R}^n \times \mathbb{R}^n \to \mathbb{R}$, such that

$$P(t,x,E) =: \int_E p(t,x,y)\, dy$$

with

$$0 \leqq p(t,x,y) < \infty$$

for $t > 0$, $x,y \in \mathbb{R}^n$.

Definition 2: Feller semigroup in L^∞

Let p be a density of a process satisfying Assumption A. Then the Feller semigroup is defined by

$$(T_t f)(x) := \int_{\mathbb{R}^n} p(t,x,y)\, f(y)\, dy, \qquad t > 0,$$

and

$$(T_0 f)(x) := f(x)$$

for $f \in L^\infty(\mathbb{R}^n, \mathbb{R})$ (essentially bounded functions).

For spectral theoretical considerations it is necessary to define the Feller semigroups in $L^2(\mathbb{R}^n)$ or more general in $L^q(\mathbb{R}^n)$, $1 \leqq q \leqq \infty$.

Definition 3: Feller semigroup in L^q

Let $\{T_t,\ t \geq 0\}$ be a Feller semigroup in $C_\infty(\mathbb{R}^n)$ (continuous functions vanishing at infinity). The semigroup is said to act in $L^q(\mathbb{R}^n)$ if it has the following properties:

a) There is a $T > 0$, $M \geq 1$ such that

$$T_t \; C_0(\mathbb{R}^n) \subset L^q(\mathbb{R}^n)$$

(C_0 - space of continuous functions with compact support) and

$$\| T_t f \|_{L^q} \leq M \| f \|_{L^q} \quad , \qquad 0 \leq t \leq T .$$

b) $\lim\limits_{t \to 0} \| T_t f - f \|_{L^q} = 0$ for $f \in C_0(\mathbb{R}^n)$.

Remark 4: This definition is taken from van Casteren [1] p. 16. Because $C_0(\mathbb{R}^n)$ is dense in $L^q(\mathbb{R}^n)$ the semigroup can be extended to L^q. Here we denote this extension by the same symbol T_t.

The objective in this section is to present conditions such that the generators of the Feller semigroups considered are selfadjoint. A first step in this direction is to study symmetric semigroups. Then we restrict us to selfadjoint semigroups in $L^2(\mathbb{R}^n)$ with selfadjoint generators.

Proposition 5: Let T_t be a Feller semigroup which is symmetric, i.e.

$$\int_{\mathbb{R}^n} (T_t f)(x)\, g(x)\, dx = \int_{\mathbb{R}^n} f(x)\, (T_t g)(x)\, dx$$

for $f, g \in C_0(\mathbb{R}^n)$. Then

$$\| T_t f \|_{L^q} \leq \| f \|_{L^q} \quad , \qquad t \geq 0,$$

for $f \in C_0(\mathbb{R}^n)$ and the semigroup acts in $L^q(\mathbb{R}^n)$, $1 < q < \infty$. For $q=2$ it extends to a selfadjoint semigroup.

Proof see [1] p. 19.

Definition 6: Feller generator

Let T_t be a contractive Feller semigroup in $L^2(\mathbb{R}^n)$. Then its generator K is given by

$$K f := \lim_{t \to 0} t^{-1} (1 - T_t) f$$

with

$$\operatorname{dom} K := \{ f,\ f \in L^2(\mathbb{R}^n),\ \lim t^{-1} (1 - T_t) f \text{ exists} \} .$$

By means of K we rewrite

$$T_t f =: e^{-tK} f \quad , \quad t \geq 0 .$$

Proposition 7: The generator of a Feller semigroup in $L^2(\mathbb{R}^n)$ is selfadjoint if the semigroup is symmetric.

Proof see [1] p. 138 .

One can find sufficient conditions for the transition density function such that the corresponding Feller semigroup acts in $L^2(\mathbb{R}^n)$.

Assumption B: Let $(\Omega_x, \mathcal{L}_x, P_F, \omega(.))$ be a Feller process homogeneous in time possessing a transition density function p. In addition we assume for this density function

$$p(t,x,y) \leqq a\,(2\pi t)^{-n/2}\; e^{-b\,|x-y|^2/2t}$$

for $0 < t \leqq T$, $x \in \mathbb{R}^n$, $y \in \mathbb{R}^n$ with some positive constants a, b .

Proposition 8: The Feller semigroup of a process satisfying Assumption B is a contractive C_o-semigroup acting in $L^2(\mathbb{R}^n)$. If $p(t,x,y) = p(t,y,x)$ for $0 < t \leqq T$ this extends to a selfadjoint semigroup with a selfadjoint generator.

Proof: The contractive C_o-property was shown in [2] p. 36ff. The selfadjointness of the semigroup and the generator follows on account of the Propositions 5 and 7 . Q.e.d.

Corollary 9: Let Assumption B be satisfied. Then it holds for all $f \in L^2(\mathbb{R}^n)$, $x \in \mathbb{R}^n$

$$(e^{-tK} f)(x) = \int_{\mathbb{R}^n} f(y)\; p(t,x,y)\; dy$$

$$= \int_{\Omega_x} f(\omega(t))\; P_F(d\omega)\;.$$

Proof: The first equation follows by the Definition 6. The second equation follows by the definition of the measure $P_F(.)$. Proposition 8 implies that these equations are true for all $f \in L^2(\mathbb{R}^n)$.

Examples:

1. The main example is the Wiener process. Its density function is symmetric. Its generator is the Laplace operator in $L^2(\mathbb{R}^n)$.
2. Another examples are time- homogeneous diffusion processes with a strongly elliptic, bounded, Hölder-continuous diffusion matrix and a bounded, Hölder-continuous drift vector.
   Further examples are given in [2] .

3. Regular perturbations of Feller generators

For introducing the Feynman-Kac formalism there are two possibilities. One can start with the unperturbed semigroup e^{-tK}, uses the Trotter-product formula, and constructs the perturbed semigroup $e^{-t(K+V)}$ where V is the multiplication operator with a potential function $V(.): \mathbb{R}^n \to \mathbb{R}$. (see e.g. Theorem 3.8. in [2]).

The other possibility is to define a function

$$(P_t f)(x) = \int_{\Omega_x} \exp\left[-\int_0^t V(\omega(s))\, ds\right] f(\omega(t)) \quad P_x(d\omega) \tag{1}$$

$x \in \mathbb{R}^n$, and study then the properties of $P_t f$ in dependence of V (see e.g. [1] p. 20ff.). Here we describe the second way and define the most important class of potentials (see Theorem 11) :

<u>Definition 10:</u> Kato-class potentials

Fix a Feller-semigroup T_t, $t \geq 0$. A Borel measurable function V mapping $\mathbb{R}^n \to \mathbb{R}$ is said to be in Kato's class with respect to T_t if it satisfies the following conditions:

1) Setting $V_+ = \max(V,0)$, it holds

$$\int_0^t (T_s V_+)(x)\, ds \quad < \infty \tag{2}$$

for $x \in \mathbb{R}^n$, $t > 0$.

2) It holds

$$\lim_{t \to 0} \sup_{x \in \mathbb{R}^n} \int_0^t (T_s V_-)(x)\, ds = 0 \tag{3}$$

with $V_- = \max(-V, 0)$.

<u>Theorem 11:</u> Let V be a Kato-class potential with respect to the Feller semigroup e^{-tK} considered in $L^2(\mathbb{R}^n)$.

Then there is a closed linear operator H which extends K+V and which generates a strongly continuous positivity preserving semigroup e^{-tH}. If e^{-tK} is symmetric, then H is selfadjoint and generates a selfadjoint semigroup, represented in equation (1).

Proof see [1] p. 22ff, 122, 171ff.

In the following we identify H with K+V and e^{-tH} with $e^{-t(K+V)}$. Again we mention the following examples. If e^{-tK} is the semigroup of the Wiener process then H is a selfadjoint Schrödinger operator. More general, if e^{-tK} is the semigroup of a canonical diffusion process such that K is an extension of a second order differential operator with variable coefficients, then H extends the perturbed operator K+V. (for explicit conditions for the coefficients see [2] p. 76 ff).

Crucial for Theorem 11 are the Kato-class potentials. For Feller processes satisfying Assumption B the Kato-class can be characterized in more detail.

Corollary 12: Assume a Feller process satisfying Assumption B. Take $0 < t \leq T$. Then a potential V mapping $\mathbb{R}^n \to \mathbb{R}$, $n \geq 3$, is a Kato-class potential with respect to the corresponding Feller semigroup if

$$\int_{\mathbb{R}^n} V_+(x)\ |x-y|^{-n+2}\ dx \quad < \infty \tag{4}$$

and

$$\lim_{r \to 0}\ \sup_{x \in \mathbb{R}^n} \int_{\{y : |x-y| \leq r\}} V_-(x)\ |x-y|^{-n+2}\ dx = 0\ . \tag{5}$$

Proof: That follows immediately from the estimate in Assumption B.

An equivalent condition for potentials satisfying (5) is the following.

Proposition 13: A potential satisfies the condition in (5) if and only if

$$\lim_{r \to 0}\ \sup_{x \in \mathbb{R}^n} \int_0^r ds\ s^{-n+1} \int_{|y| \leq s} V_-(x+y)\ dy = 0\ . \tag{6}$$

Proof: Integration by parts yields

$$\begin{aligned} &\int_{|x-y| \leq r} V_-(y)\ |x-y|^{-n+2}\ dy \\ &= r^{-n+2} \int_{|u| \leq r} V_-(x+u)\ du \\ &\quad + (n-2) \int_0^r ds\ \ s^{-n+1} \int_{|u| \leq s} V_-(x+u)\ du. \end{aligned} \tag{7}$$

From (5) follows

$$\sup_{x \in \mathbb{R}^n} \int_{|u| \leq r} V_-(x+u)\ du = o(r^{n-2})$$

as $r \to 0$. The same follows from the condition in (6). Thus in both cases the first term in (7) vanishes as $r \to 0$. Q.e.d.

Remark 14: The condition in (6) is satisfied for $V_- \in L^q(\mathbb{R}^n)$, $n/2 < q \leq \infty$. That implies for instance $V_- \in L^2(\mathbb{R}^3) + L^\infty(\mathbb{R}^3)$ such that Coulomb potentials are included. A necessary condition for (5) is

$$\int_{|x| \leq r} V_-(x) \quad dx = o(r^{n-2})$$

as $r \to 0$.

Introducing the conditional measure of the processes considered the Feller semigroups turned out to be integral operators.

Denotation 15: Let $(\Omega_x, \mathcal{B}_x, P_F, \omega(.))$ be the Feller process given in Assumption A. Then we set

$$\Omega_x^{y,t} := \{ \omega : \omega \in \Omega_x, \omega(t) = y \} ,$$

and we denote with $P_F^{y,t}(.)$ the conditional measure corresponding to this process.

Theorem 16: Let e^{-tK} be a Feller semigroup of a process satisfying Assumption A (see the Definitions 3 and 6). Its density function is assumed to be symmetric. Let V be a Borel measurable function from $\mathbb{R}^n$ to $\mathbb{R}$ which belongs to Kato's class.

Then the semigroup $e^{-t(K+V)}$ consists of integral operators the kernels of which are given by

$$(e^{-t(K+V)}) \ (x,y) := \int_{\Omega_x^{y,t}} \exp\left[- \int_0^t V(\omega(s) \ ds \right] P_F^{y,t}(d\omega) \quad . \tag{8}$$

The kernels are continuous on $(0,\infty) \times \mathbb{R}^n \times \mathbb{R}^n$ and symmetric.

Proof: The proof is given in [1] p. 35ff using essentially Khas'minskii's Lemma. Another proof is given in [2] p. 35ff for potentials which can be approximated by bounded ones.

Corollary 17: If V is a bounded measurable function from $\mathbb{R}^n \to \mathbb{R}$ it holds

$$(e^{-t(K+V)})(x,y) = e^{t \ \text{ess} \sup_{x \in \mathbb{R}^n} |V_-(x)|} \ p(t,x,y)$$

where p is the density function of the Feller process considered.

Proof: This follows immediately from the representation in (8).

Besides this trivial estimation of the kernel a more general one is given recently by van Casteren in [4] .

Proposition 18: Let K be the Laplace operator and assume a Kato-class potential V, i.e. V shall satisfy the conditions in (4) and (5). Then there are positive constants α and β such that

$$(e^{-t(K+V)})(x,y) \leqq \alpha\,(2\pi t)^{-n/2}\; e^{\beta t}\; e^{-|x-y|^2/2t}$$

for all $t > 0$.

4. Singular perturbations of Feller generators

In this section we neglect the regular perturbations for the sake of simplicity, i.e. we assume $V(x) \equiv 0$, but we consider singular perturbations of Feller generators which arise by increasing potential barriers up to infinity. This region over which the potential barriers are defined is called singularity region. Describing that we have to introduce some further denotations.

Definition 19: Singularity region

Let G be a closed region in $\mathbb{R}^n$ with a piecewise $\mathcal{C}^1$ boundary. Let $\chi_G(\cdot)$ be the corresponding indicator function and P be a projection operator in $L^2(\mathbb{R}^n)$ given by

$$(Pf)(x) := \chi_C(x)\; f(x) \quad , \qquad f \in L^2(\mathbb{R}^n) .$$

Moreover, we denote with $\bar{P}$ the complementary projection operator with respect to P, i.e.

$$\bar{P} := 1 - P .$$

With J^* we denote the canonical embedding operator from $L^2(\mathbb{R}^n \setminus G)$ to $L^2(\mathbb{R}^n)$, i.e.

$$J^* := 1_{L^2(\mathbb{R}^n \setminus G) \to L^2(\mathbb{R}^n)} \quad .$$

Let J be the adjoint of it such that

$$J^* J = 1_{L^2(\mathbb{R}^n)} - P = \bar{P} \quad , \qquad .$$

$$J J^* = 1_{L^2(\mathbb{R}^n \setminus G)} \quad .$$

As mentioned above, singular perturbations of Feller generators are introduced by increasing potentials over the singularity region G .

Theorem 20: Assume a Feller process which satisfies Assumption B. Let e^{-tK} be the corresponding semigroup with the Feller generator K considered in $L^2(\mathbb{R}^n)$. Moreover, assume a positive parameter M and define

$$V_M := M\, P$$

which is given by the multiplication operator

$$V_M(\cdot) = M\, \chi_G(\cdot) \quad .$$

Set $K_M := K+V_M$. Then $\operatorname{dom} K_M = \operatorname{dom} K$ and K_M generates a contractive C_0-semigroup, e^{-tK_M} , in $L^2(\mathbb{R}^n)$.

Then this semigroup has a strong limit as $M\to\infty$, which is denoted with U_t , i.e.

$$\lim_{M\to\infty} \| e^{-tK_M} f - U_t f \| = 0 , \qquad t > 0, \tag{9}$$

for all $f\in L^2(\mathbb{R}^n)$. The limit U_t can be represented pointwise by

$$(U_t f)(x) = \int_{\Omega_x} \chi\{\omega : \omega(s) \notin G, \forall s,\ s\in[0,t]\}\ f(\omega(t))\ P_F(d\omega) \tag{10}$$

for a.a. $x\in\mathbb{R}^n\setminus G$ and all $f\in L^2(\mathbb{R}^n)$.

Moreover $U_t \restriction L^2(\mathbb{R}^n\setminus G)$ is a contractive C_0 semigroup. Denoting its generator with $(K)_G$ this semigroup consists of integral operators, the kernels of which are given by

$$(e^{-t(K)_G})\ (x,y)$$

$$= \int_{\Omega_x^{y,t}} \chi\{\omega : \omega(s) \notin G,\ \forall s,\ s\in[0,t]\}\quad P_F^{y,t}(d\ ) \quad . \tag{11}$$

Proof: The proof is given in [2] p. 43ff and 51ff.

In the last section the selfadjointness of the generators K and $K+V$ was studied. In dependence of K also the singularly perturbed generator $(K)_G$ can be described.

Theorem 21: Let K be a selfadjoint, nonnegative operator. Let $\bar{P}$ be given as in Definition 19. Suppose that $L^2(\mathbb{R}^n \setminus G) \cap \mathrm{dom}\, K$ is dense in $L^2(\mathbb{R}^n \setminus G)$, and assume $Kf = \bar{P}Kf$ for all $f \in L^2(\mathbb{R}^n \setminus G) \cap \mathrm{dom}\, K$. Set again $K_M = K + MP$, $M>0$, and choose $z \in \bigcap_M \mathrm{res}\, K_M$ (resolvent sets).

Then the strong limit of $(z - K_M)^{-1}$ exists, i.e.

$$\operatorname*{s-lim}_{M \to \infty} (z - K_M)^{-1} =: R_G(z) ,$$

where $R_G(z)$ is a pseudo-resolvent. Moreover, $R_G(z) \restriction L^2(\mathbb{R}^n \setminus G)$ is the resolvent of $(K)_G$, and $(K)_G$ is the Friedrichs-extension of $K \restriction [\mathrm{dom}\, K \cap L^2(\mathbb{R}^n \setminus G)]$.

Proof: The proof is given by Baumgärtel, Demuth in [5] .

Remark 22: Regular perturbations could be included into the considerations of this and the next section by using the estimations of the form given in Corollary 17 and Proposition 18.

5. Spectral properties of Feller generators

Here the influence of the singular perturbations for the spectrum of Feller generators is discussed by investigating the differences (see Definition 19)

$$e^{-t(K)_G} J - J e^{-tK}$$

of Feller semigroups mapping $L^2(\mathbb{R}^n) \to L^2(\mathbb{R}^n \setminus G)$. The denotations are taken from the sections above.

Theorem 23: Assume that e^{-tK} is a Feller semigroup of a Feller process which satisfies Assumption B. With p we denote again the density of the process.

Then the semigroup difference

$$e^{-t(K)_G} J - J e^{-tK}$$

is a Hilbert- Schmidt operator for any $t \in (0,T]$ if

$$\sup_{s \in (0,t)} \int_G p(s,x,u)\, du \in L^1(\mathbb{R}^n \setminus G, dx).$$

(Compare this with the Kato-class conditions for the regular potential parts in Definition 10) .

Proof: $e^{-t(K)_G} J - J e^{-tK}$ is an integral operator. Because of the representations in Corollary 9 and Theorem 20 the following estimations are valid:

$$\int_{\mathbb{R}^n\setminus G} dx \int_{\mathbb{R}^n} dy \; | (e^{-t(K)_G} J - J e^{-tK})(x,y)|^2$$

$$= \int_{\mathbb{R}^n\setminus G} dx \int_{\mathbb{R}^n} dy \; | \int_{\Omega_x^{y,t}} \chi\{\omega : \exists s_o,\ s_o \in (0,t),\ \omega(s_o) \in G\} \, P_F^{y,t}(d\omega)|^2$$

$$\leq \text{const} \cdot \int_{\mathbb{R}^n\setminus G} dx \int_{\Omega_x} \chi\{\omega : \exists s_o,\ \omega(s_o) \in G\} \, P_F(d\omega) \;.$$

Now let

$$\theta_m = \{\lambda_o = 0,\ \lambda_1,\ \dots ,\ \lambda_{m-1},\ \lambda_m = t\} \quad , \ \lambda_i < \lambda_{i+1} ,$$

be a partition of $[0,t]$. The trajectories are continuous. The measure of the trajectories which only touch the boundary of G is zero (see Ginibre [6]) . Therefore

$$\chi\{\omega : \exists s_o,\ s_o \in (0,t),\ \omega(s_o) \in G\}$$

$$= \lim_{m\to\infty} \chi\{\omega : \exists \tau_o ,\ \tau_o \in \theta_m ,\ \omega(\tau_o) \in G\} \;.$$

But the last expression is smaller than

$$\sup_{s\in(0,t)} \int_{\mathbb{R}^n} p(s,x,u) \, \chi_G(u) \, du \;. \qquad \text{Q.e.d.}$$

<u>Theorem 24:</u> Let the assumptions of the Theorem 23 be satisfied. Then the semigroup difference

$$e^{-t(K)_G} J - J e^{-tK} , \qquad 0 < t = T ,$$

is a trace class operator , if

$$(1+|x|)^\delta \sup_{s\in(0,t)} \int_G p(s,x,u) \, du \in L^1(\mathbb{R}^n \setminus G,\ dx) \tag{12}$$

with $\delta > n$.

Proof: Let Q be the multiplication operator, given by

$$(Qf)(x) := (1 + |x|)^{\delta/2} f(x) ,$$

$x \in \mathbb{R}^n$, $\delta > n$, $f \in L^2(\mathbb{R}^n)$. Then we rewrite the semigroup difference

$$e^{-2t(K)_G} J - J e^{-2tK}$$

$$= e^{-t(K)_G} J Q^{-1} \; Q J^* (e^{-t(K)_G} J - J e^{-tK})$$

$$+ (e^{-t(K)_G} J - J e^{-tK})Q \; Q^{-1} e^{-tK} \quad .$$

Because of Assumption B the operators $e^{-t(K)_G} J Q^{-1}$ and $Q^{-1} e^{-tK}$ are Hilbert-Schmidt operators. Corresponding to the proof of Theorem 23 the operators $Q J^* (e^{-t(K)_G} J - J e^{-tK})$ and $(e^{-t(K)_G} J - J e^{-tK})Q$ are also Hilbert-Schmidt operators, if the condition in (12) is fulfilled. Q.e.d.

<u>Consequences 25:</u> If one has selfadjoint Feller generators K and also selfadjoint $(K)_G$, then the Hilbert-Schmidt property in Theorem 23 implies a stable essential spectrum, i.e.

$$\sigma_{ess}(K) = \sigma_{ess}((K)_G) \quad .$$

That follows by a general two-space criterion given by Brüning, Demuth, Gesztesy in [7] .

If the trace class condition in Theorem 24 is satisfied one can use the invariance principle of the mathematical scattering theory to obtain for the absolutely continuous spectra

$$\sigma_{ac}(K) = \sigma_{ac}((K)_G) \quad .$$

The abstract condition in Theorem 24 admits unbounded singularity regions. Here we apply it to star-like regions. A star is a union of peaks. Thus it suffices to consider a single peak of such a star.

<u>Definition 26:</u> Peak of a star

For the sake of simplicity we assume that the peak D is unbounded in the direction of the x_1-axis. Let $x \in \mathbb{R}^n$ be decomposed into $(x_1, \vec{x}_{n-1})$, $x_1 \in \mathbb{R}$, $\vec{x}_{n-1} \in \mathbb{R}^{n-1}$. $\vec{x}_{n-1}$ shall be orthogonal to the x_1-axis. With P_{x_1} we denote the plane through $(x_1, \vec{0})$ orthogonal to the x_1-axis. Let $D(x_1)$ be the orthogonal projection of $P_{x_1} \cap D$ onto P_0. $|D(x_1)|$ is the (n-1)-dimensional volume of $D(x_1)$.

Then D is assumed to satisfy the following conditions:

- D is connected and closed.
- For all $x_1 \geq x_1' \geq 0$ and for all $x_1 \leq x_1' \leq 0$ it holds $D(x_1) \subseteq D(x_1')$.
- $D(0)$ is bounded in $\mathbb{R}^{n-1}$.
- The boundary of D is piecewise in $\mathcal{C}^1$. The boundary can be estimated by f_j, g_j, $j=2,\ldots,n$, mapping $\mathbb{R}^{j-1}$ into $\mathbb{R}$ such that

$$D(x_1) := \{ \vec{x}_{n-1} = (x_2,\ldots,x_n) :$$
$$f_j(x_1,\ldots,x_{j-1}) \leq x_j \leq g_j(x_1,\ldots,x_{j-1}),\ j=2,\ldots,n \}.$$

Because $D(x_1)$ is contained in $D(0)$ f_j and g_j are bounded functions. Therefore we can define

$$d := 2 \sup_{j=\{2,\ldots,n\}} \sup_{\vec{x}_{j-1} \in \mathbb{R}^{j-1}} (|f_j(\vec{x}_{j-1})| , |g_j(\vec{x}_{j-1})|).$$

<u>Theorem 27:</u> Assume a Feller process which satisfies Assumption B. Let D be a peak of a star according to the Definition 26.

Then the semigroup difference

$$e^{-t(K)_D} J - J e^{-tK} \qquad , 0 < t \leq T ,$$

is a trace class operator, if the peak D satisfies

$$\left| \int_{-\infty}^{\infty} |x_1|^{\alpha} |D(x_1)| \, dx_1 \right| < \infty$$

with $\alpha > n$.

Note that unbounded D are admitted. But the amount of the contraction of the peak must have a sufficiently high power in $|x_1|^{-1}$ as $|x_1| \to \infty$.

Proof: The proof ideas are sketched. One has to prove that the condition in Theorem 24 is satisfied. The main difficulties arises from such trajectories starting near the boundary of D. Therefore one introduces a larger region $\tilde{D}$, $\tilde{D} \supset D$, the boundary of which has a decreasing distance to δD as $|x_1| \to \infty$. Set

$$\tilde{D}(x_1) := \{ \vec{x}_{n-1} : f_j - (1+|x_1|)^{-\mu} \leqq x_j \leqq g_j + (1+|x_1|)^{-\mu}; \\ j = 2,\ldots,n \}$$

with $\mu > \delta > n$. The volume of $\tilde{D}$ is finite. Hence it suffices to estimate

$$\sup_{s \in (0,t)} \int_D p(s,x,u)\, du$$

for $x \in \mathbb{R}^n \setminus \tilde{D}$. Because of Assumption B this is smaller than

$$c \cdot \int_D |x-u|^{-n} \quad e^{-c|x-u|^2} \quad du$$

(c is some constant). For $\vec{x}_{n-1} \in \mathbb{R}^{n-1} \setminus \tilde{D}(x_1)$ and $\vec{u}_{n-1} \in D(u_1)$ the definition of $\tilde{D}$ provides

$$|\vec{x}_{n-1} - \vec{u}_{n-1}| \geqq (1 + |x_1|)^{-\mu}.$$

This estimation is used to prove that the integral

$$\int_{\mathbb{R}^n \setminus \tilde{D}} dx \ (1+|x|)^{\delta} \int_D du \ |x-u|^{-n} \quad e^{-c|x-u|^2} \quad du$$

is finite if

$$\int_{-\infty}^{\infty} |x_1|^{\delta+\varepsilon} \ |D(x_1)| \ dx_1 < \infty \quad , \quad \varepsilon > 0 \ . \qquad \text{Q.e.d.}$$

References

1. J. van Casteren: Generators of strongly continuous semigroups, Pitman; Boston, London, Melbourne, 1985.

2. M. Demuth, Report Inst. Math. Berlin, 1985, R-Math-04/85.

3. A. Friedman: Stochastic differential equations and applications, Vol.1. ; Academic Press; New York, San Francisco, London,1975.

4. J. van Casteren, Integral kernels for Schrödinger type equations, Preprint Univ. Antwerpen, 1986, nr.86-32 .

5. H. Baumgärtel, M. Demuth, Rep. Math. Phys. 1979, v.15, p.173 .

6. J. Ginibre: Some applications of functional integration in statistical mechanics and quantum field theory. - In: Statistical mechanics and quantum field theory, Les Houches 1970, ed. by C. DeWitt, R. Stora; Gordon and Breach; New York, London, Paris; 1971, p. 327 .

7. E. Brüning, M. Demuth, F. Gesztesy, Lett. Math. Phys. 1987, v. 13, p. 69 .

SINGULAR PERTURBATIONS DEFINED BY FORMS

V.D.Koshmanenko

Institute of Mathematics, Ukrainian Acad. of Sci.

252601, Kiev, USSR

We give a precise sense to the notion of singular perturbation. It is a bilinear form b in a Hilbert space H with a regular (closable) component $b_r = o$. Further we propose a classification of singular bilinear forms with respect to a fixed selfadjoint operator $A \geqslant o$ in H. Finally we present a construction of the singularly perturbed operator A_b. Our definition of A_b is based on the interpretation of b as a boundary condition for a fixed selfadjoint extension of the symmetric operator $A_o = A \restriction \mathrm{Ker}\ b$.

1. Introduction

A singular perturbation of a selfadjoint operator A in Hilbert space H is a formal mathematical expression given on a linear set $G \subseteq D(A)$ and which is equal to zero on a subset $G_o \subset G$ supposed to be dense in H. As a rule such expressions can be given a sense as bilinear forms in H. But these forms are not closable, and therefore they have not an operator representation in H. In physics such expressions are used for heuristic description of point, relativistic-point and other types interactions. All these interactions are concentrated within a "small" space-time region (with Lebesgue measure zero) which is the reason for the lack of correct operator sense for these so-called singular interactions. In spite of zero measure of the interaction regions many of them represent nontrivial perturbations.

At the present time, there is an increasing number of publications dealing mainly with point interactions (see, for example, papers by S.Albeverio, R.Hoegh-Krohn and others [1-4]). The key problem is to give a reasonable interpretation to the Schrödinger operator with a point interaction. In general, there are several approaches to this problem. But a complete and consistent theory of singular perturbations or point interaction alone has not been formulated up to now.

In the present paper, we are going to develop the following point of view. A singular perturbation can be interpreted in terms of a boundary condition corresponding to a fixed selfadjoint extension of some Hermitian operator which is common for both the free and the singularly perturbed Hamiltonians. For a wide class of singular perturbations, we formulate this interpretation in a mathematically correct way.

The idea to make use of the selfadjoint-extension theory to study the formal Schrödinger operator $A_\delta = -\Delta + \delta(x)$ belongs to Berezin and Faddeev [5]. They proposed to interpret A_δ as the family of all selfadjoint extensions of the Hermitian operator $A_o = -\Delta \upharpoonright G_o$, where $G_o \subset D(-\Delta)$ consists of the functions $u(x)$ such that $u(o) = o$. The question about choise of a particular selfadjoint extension in general is open. We are going to show that in the general case there is a one-to-one correspondence between some class of singular perturbations and the set of all selfadjoint extensions of a symmetric operator which is analogous to A_o. To formulate the correspondence rule, we treat the singular perturbations in a non-additive way. The information about a perturbation is contained in boundary conditions of some new (different from the free one) selfadjoint operator.

2. Definition of a singular perturbation

Let G be a linear dense subset in a complex Hilbert space H. A map $b: G \times G \to \mathbb{C}$ such that $b(u,v)$ is linear in $u \in G$ and conjugate linear in $v \in G$ is called a bilinear form on G with the domain $Q(b) = G$. The inner product $(\cdot,\cdot)$ in H restricted to G gives an example of a bilinear form on G. It will be denoted in the following as a. A form b is possitive if $b[u] \equiv b(u,u) \geqslant 0$ for all $u \in G$.

Definition 2.1: A positive bilinear form b on G is singular in H (notation $b \perp a$) if for earh $u \in G$, there exists a sequence $\{u_n\} \subset G$, $n = 1,2, \ldots$ such that

$$u_n \xrightarrow{H} u, \quad b[u_n] \to 0, \quad n \to \infty . \tag{2.1}$$

We write $b \in F_s$ if $b \perp a$.

Let H_b denote the Hilbert space which is obtained by completition of G equipped with the (quasi)inner product $(u,v)_b = b(u,v)$. In particular $H_a \equiv H$. The fact that b is singular in H means that H_b and H_a are orthogonal in following sense.

Proposition 2.1 [6] :

$$b \perp a \Longleftrightarrow H_{a+b} = H_a \oplus H_b \quad . \tag{2.2}$$

Note that (2.2) implies $b \perp a \Longleftrightarrow a \perp b$. It means that for all $v \in G \subset H_b$, there exists a sequence $\{v_n\} \subset G$ such that

$$v_n \xrightarrow{H_b} v, \quad v_n \xrightarrow{H} 0, \quad n \to \infty . \tag{2.1'}$$

It is easy to see that u and v in (2.1) and (2.1') can be arbitrary vectors from H and H_b.

The condition (2.1) is fulfilled automatically if the set

$$G_o \equiv \text{Ker } b = \{ u \in G \mid b[u] = o \}$$

is dense in H. In what follows we study the singular forms for which G_o is dense in H, i.e.

$$\overline{G}_o = H. \tag{2.3}$$

<u>Example</u>: Let $H = L_2(R^d)$, $d = 1,2, \ldots$, $A = -\Delta$ and $G = S(R^d)$ be the Schwarts space. The bilinear form

$$b_o(u,v) = u(o)\,\overline{v(o)}, \quad u, \quad v \in S(R^d)$$

is singular in L_2 since $G_o = \{ u \in S(R^d) \mid u(o) = o \}$ is dense in L_2.

Next we introduce a class of singular bilinear forms of rank one. Let $H_- \supset H \supset H_+$ be a rigged Hilbert space; $H_+ \supset G$. We ascribe to each vector $w \in H_-$ the bilinear form b_w on G defined by the formula

$$b_w(u,v) = \langle u,w \rangle \langle w,v \rangle ,$$

where $\langle \cdot , \cdot \rangle$ is the duality between H_+ and H_-.

<u>Proposition</u> 2.2: The form b_w is singular in H iff $w \in H_- \setminus H$.

<u>Proof</u>. Let $G_o = \text{Ker } b_w$. Since b_w is a form of rank one, the equality (2.3) is fulfilled only if $w \in H_- \setminus H$.

Consider a negative Sobolev space $H_- = W_2^{-k}(R^d)$, $k = 1,2\ldots$. Then to each distribution $w \in W_2^{-k} \setminus L_2$ the above defined singular form b_w in L_2 corresponds. In particular if $k > d/2$, then $\delta \in W_2^{-k}$ and for $w = \delta$ we have $b_w = b_\delta$.

Let A be a positive selfadjoint operator H, and consider a set G such that

$$G \subset D(A), \quad AG \subset G, \quad \overline{A \upharpoonright G} = A. \tag{2.4}$$

Each bilinear form b on G such that the set $G_o = \mathrm{Ker}\, b$ is dense in H will be called the singular perturbation of A. Our aim is to construct the corresponding perturbed operator which we denote as A_b. Before explaining our construction of A_b, we give a classification of the singular perturbations introduced above.

For this purpose we introduce the A-scale of Hilbert spaces:

$$\ldots H_{-k} \supset \ldots H_{-1} \supset H_o \equiv H \supset H_1 \supset \ldots H_k \supset \ldots G, \tag{2.5}$$

where $H_{\pm k}$, $k = o,1, \ldots$ is the completion of G with respect to the inner products

$$(u,v)_{\pm k} \quad a_{\pm k}(u,v) \quad ((A+1)^{\pm k}u,v), \; u, \, v \in G. \tag{2.6}$$

<u>Definition</u> 2.2: The bilinear form b on G is k-singular ($k = o,1, \ldots$) with respect to A (notation $b \perp a_k$) if b is singular in H_k but it is closable in H_{k+1}. The set of all k-singular forms we denote as F_{s_k}.

<u>Theorem</u> 2.1 [7] : For each positive bilinear form b in H with $Q(b) = G \subset D(A)$ the following decomposition is valid

$$b = b_r + b_{s_o} + b_{s_1} + \ldots, \quad b_{s_k} \in F_{s_k}, \quad k=o,1, \ldots$$

where $b_r \subseteq \overline{b}_r$ is the largest part of b which is closable in H. All components b_r, b_{s_k}, $k = o,1, \ldots$, are uniquely determined by b.

Note that the decomposition of a bilinear form b on the regular b_r and the singular $b_s = b - b_r$ parts has first appeared in [8] (see also [6]).

Let b be an arbitrary positive form on G. When b belongs to F_{s_k} ? Assume that b is bounded or closable in H_{k+1} for some fixed k : $b \subseteq \bar{b}^{(k+1)}$. Then a selfadjoint operator V_b in H_{k+1} is associated with $b^{(k+1)}$ due to the representation theorem

$$\bar{b}^{(k+1)}(u,v) = (V_b\, u,\, v)_{k+1},\ u,v \in D(V_b) \subset Q(\bar{b}^{(k+1)}) .$$

We can also associate with $\bar{b}^{(k+1)}$ the operator

$$T_b = I_{k+1}^{-1}\, V_b\colon\ H_{k+1} \to H_{-(k+1)}$$

where $I_{k+1}\ :\ H_{-(k+1)} \to H_{k+1}$ is the unitary operator which is obtained as the closure of the map

$$H_{-(k+1)} \ni u \to (A+1)^{-(k+1)}\ u \in H_{k+1},\quad u \in G.$$

Thus

$$b(u,v) = \langle T_b u, v\rangle \quad ,\quad u,v \in D(V_b) = D(T_b) \subset H_{k+1}.$$

Let us decompose H_{k+1} into two subspaces $H_{k+1} = K_b \oplus R_b$, where $K_b \equiv \operatorname{Ker} V_b = \bar{G}_o^{(k+1)}$ and $R_b = K_b^{\perp}$. The restriction of $\bar{b}^{(k+1)}$ to R_b we denote as $b^{\wedge}$.

<u>Theorem</u> 2.2: Let $b^{\wedge} \geqslant m_b > 0$ in R_b. Then the following three conditions are equivalent:

(i) $b \in F_{s_k}$ $\quad (b \perp a_k)$,

(ii) $\operatorname{Ran}(T_b) \subset H_{-(k+1)} \setminus H_{-k}$,

(iii) K_b is dense in H_k .

Proof. Let $b \in F_{s_k}$, i.e., $b \perp a_k$. Assume that (ii) is not valid, i.e., that for some $u \in D(T_b)$ the vector $w = T_b u$ belongs to H_{-k} and $w \neq o$. Then $\bar{b}^{(k+1)}[u] > o$ and since $b \perp a_k$, there exists a sequence $\{u_n\} \subset G$ such that $a_k[u_n] \equiv \|u_n\|_k^2 \to o$ and $\bar{b}^{(k+1)}[u_n - u] \to o$, $n \to \infty$.

This fact means that $\bar{b}^{(k+1)}[u_n] \to -b^{(k+1)}[u] \neq 0$

since $|\bar{b}^{(k+1)}(u_n, u)| = |\langle u_n, T_b u\rangle| = |\langle u_n, w\rangle| \leqslant \|u_n\|_k \ \|w\|_{-k} \to o$, but this is impossible if $b \geqslant o$. Hence (i) $\Longrightarrow$ (ii). Let now $\mathrm{Ran}(T_b) \cap H_{-k} = \{o\}$. Assume that (iii) is not valid, i.e., that for a non-zero vector $v \in H_k$, it holds $(K_b, v)_k = o$. It means that $I_{k+1} I_k^{-1} v \in R_b = \mathrm{Ran}(V_b)$, where the last equality follows from $b \restriction \geqslant m_b > o$. Hence $I_k^{-1} v \in \mathrm{Ran}(T_b)$, but this is impossible if $I_k^{-1} v \in H_{-k}$. Thus (ii) $\Longrightarrow$ (iii). The implication (iii) $\Longrightarrow$ (i) is equivalent to that (2.3) $\Longrightarrow$ (2.1).

3. Construction of the singularly perturbed operator

Let $A = A^* \geqslant o$ and $b \in F_s$ with $Q(b) = G$ which satisfies (2.4). Assume that $G_o \equiv \mathrm{Ker}\, b$ is dense in H. The problem is to ascribe to a pair A, b a unique selfadjoint operator A_b which is interpreted the singularly perturbed operator. We start with some preliminaries.

Denote $A_o = \overline{A \restriction G_o}$. The operator A_o is symmetric. What are its deficiency indices $n^{\pm}(A_o)$?

Theorem 3.1 [7]: Suppose that b belongs to F_{s_0} or F_{s_1}, and $b \restriction \geqslant m_b > o$. Then

$$n^{\pm}(A_o) = \dim H_b = \dim \mathrm{Ran}(V_b) = \dim \mathrm{Ran}(T_b). \tag{3.1}$$

If $b \in F_{s_k}$, $k \geqslant 2$, the deficiency indices are zero, G_o is dense in H_2 and A_o is essentially selfadjoint. In addition, A is different from the Friedrichs extension A_∞ of A_o if $k = o$, while $A = A_\infty$ for $k = 1$.

In particular, if $b = b_w$, where $w \in H_{-2} \setminus H$, then the numbers $n^{\pm}(A_o) = 1$ because $\dim H_{b_w} = 1$. On the other hand, if $w \in H_{-k} \setminus H_{-2}$, $k > 2$, (i.e., $b_w \in F_{s_{k-1}}$), then $n^{\pm}(A_o) = o$ and $A_o = A$.

Thus all forms $b \in F_{s_k}$ with $k \geqslant 2$ are the trivial perturbations of A.

Let $b \in F_{s_o}$ or F_{s_1} and $b^{\upharpoonright} \geqslant m_b > o$ in H_2. It is convenient to put $A \geqslant m > o$. Then the A - scale (2.5) is determined equivalently by the inner product $(u,v)_{\pm k} = (A^{\pm k}u,v)$ instead of (2.6). Denote the range of A_o by M and put $N = \operatorname{Ker} A_o^*$ so that $H = M \oplus N$. The space H_2 decomposes naturally into the orthogonal sum $H_2 = K_b \oplus R_b$, where $K_b = \overline{G}_o^{(2)}$ and $R_b = \operatorname{Ran}(V_b)$. The following equality is crucial for construction of A_b.

<u>Proposition 3.1:</u>

$$AR_b = N, \tag{3.2}$$

where $A = I^{-1} : H_2 \to H$ is the unitary operator.

Proof. We have $(AR_b, M) = (AR_b, \overline{AG_o}) = (R_b, K_b)_{+2} = 0$ which means that $AR_b \subseteq N$. On the other hand, $(A^{-1}N, K_b)_{+2} = (AA^{-1}N, AK_b) = (N, M) = o$, i.e., $A^{-1} N \subseteq R_b$.

Now we can associate with b the operator B_b in N defined by

$$B_b = A\, V_b^{\upharpoonright} A^{-1} \equiv I^{-1}\, V_b^{\upharpoonright} I\,. \tag{3.3}$$

Since $b^{\wedge} \geqslant m_b > o$, the operator B_b is positive and invertible. For construction of A_b, we need the following well-known abstract result ([9, 10] ; see also [11, 12]).

<u>Theorem</u> 3.2: Let $A_o \geqslant m > o$ be a closed Hermitian operator such that $\operatorname{Ker} A_o^* \equiv N \neq o$. Then each positive selfadjoint extension $\widetilde{A}$ of A_o is given by the formula

$$\widetilde{A}^{-1} = A_\infty^{-1} + B^{-1}, \tag{3.4}$$

where $B = B^* \geqslant o$ acts in N. The correspondence between $\widetilde{A}$ and B is bijective.

This theorem establishes a parametrization of the set of all positive selfadjoint extensions of A_o in terms of auxiliary operators $B = B^* \geqslant o$ in N. In really, these operators play a role of the boundary conditions for extension $\widetilde{A}$. Due to (3.2) and (3.3) this is just the parametrization by described above singular bilinear forms. If we put $A_b \equiv \widetilde{A}$, then (3.4) implies

$$A_b^{-1} = A_\infty^{-1} + B_b^{-1} \tag{3.5}$$

The singularly perturbed operator A_b is specified by A and b if b belongs to F_{s_o} or F_{s_1}.

Next we want to describe the domain of A_b. The vector g belongs to $D(A_b)$ if it has the following representation

$$g = f + B_b^{-1} P_N A_\infty f, \qquad f \in D(A_\infty), \tag{3.6}$$

where P_N is an orthogonal projection with the range N. In addition, we have

$$A_b\, g = A_\infty f. \tag{3.7}$$

Note that (3.6) and (3.7) is true if $b^{\restriction} \geqslant m_b > o$ in H_2 and the domains fulfil $D(A_\infty^{-1}) = D(A_b^{-1}) = H$. A general description of $D(\widetilde{A})$ has been obtained in [12] .

Each operator $\widetilde{A}$ appearing in (3.4) is associated with the form

$$\widetilde{b} \equiv b_A = b_{A_\infty} \dotplus b_B , \quad Q(\widetilde{b}) = Q(b_{A_\infty}) \dotplus Q(b_B)$$

where $\dotplus$ means the direct sum; in other words $h \in Q(\widetilde{b})$ if $h = u + v$ for $u \in Q(b_{A_\infty})$, $v \in Q(b_B) \subset N$ and $\widetilde{b}[h] = b_{A_\infty}[u] + b_B[v]$ (here b_{A_∞} is the form corresponding to A_∞ and b_B corresponds to B). Vice versa the form b_B corresponding to $B \equiv B_b$, where b belongs to F_{s_0} or F_{s_1}, may be expressed directly by means of b and A:

$$b_B[v] = (B_b v, v) = (AV_b^{\restriction} A^{-1} v, v) = (AV_b^{\restriction} A^{-1} v, AA^{-1} v)$$

$$= (V_b^{\restriction} A^{-1} v, A^{-1} v)_{+2} = \bar{b}^{(2)}[A^{-1} v], \quad A^{-1} v \in Q(\bar{b}^{(2)}).$$

The operator A_b is therefore associated with the direct form sum:

$$b_{A_b}[h] = b_{A_\infty}[u] + \bar{b}^{(2)}[A^{-1} v], \; h = u \dotplus v. \qquad (3.8)$$

This decomposition is correct since $Q(b_{A_\infty}) \cap N = \{o\}$ and $N \subset \text{Ran}(A)$.

We have noticed already that A is identical with the Friedrichs extension A_∞ only if $b \in F_{s_1}$. In the case $b \in F_{s_0}$, we can define the perturbed operator using the usual form sum in view of the following fact

<u>Theorem</u> 3.3 [7]: The bilinear form $b_{A_1} \equiv b_A + b$ is closable in H iff $b \in F_{s_0}$.

The operator A_1 associated with closure of $b_A + b$ is different from A_b defined by (3.5) or (3.8). In other words, the form - sum and selfadjoint - extensions methods yield different results in the case $b \in F_{s_0}$. On the other hand, the form-sum method is not applicable for $b \in F_{s_1}$, since $b_A + b$ is not closable in this case. In addition, we have the following abstract negative result.

<u>Theorem</u> 3.4 [13] : Let a sequence of forms $\{b_n\} \subset F_{s_0}$, $n = 1,2 \ldots$ converges to $b \in F_{s_1}$ in the following sense:

$$V_{b_n} \xrightarrow{s.r.s} V_b \quad , \quad K_{b_n} \xrightarrow{s} K_b, \quad n \to \infty . \tag{3.9}$$

Then the sequence of operators A_n which are associated with $b_A + b_n$ converges to the unperturbed operator, $A_n \to A$ in the strong resolvent sense.

The replacement of A_n by A_{b_n} means a regularization, which leads to a non-trivial effect:

<u>Theorem</u> 3.5 [13] : Under the assumptions of theorem 3.4, the replacement of A_n by A_{b_n} gives $A_{b_n} \to A_b$ in strong resolvent sense, where A_{b_n} and A_b are defined by A, b_n and b according to (3.5) and (3.8).

It means that the dependence of A_b on b is continuous in the strong resolvent sense for both A_b in H and V_b in H_2.

References

1 S. Albeverio, J.E.Fenstad and R.Høegh-Krohn, Singular perturbations and nonstandard analysis, Trans. Am. Math. Soc. 252, 275-295 (1979).

2 S.Albeverio, F.Gesztesy, R.Høegh-Krohn, W.Kirsch, On point interactions in one dimension, J. Oper. Th. 12, 101-126 (1984).

3 A.Grossmann, R.Høegh-Krohn, M.Mebkhout, The one-particle theory of periodic point interactions, Commun. Math. Phys. 77, 87-110 (1980)

4 P. Šeba, Some remarks on the δ'-interaction in one dimension, Rep. Math. Phys. 24, 111-120 (1986).

5 F.A.Berezin, L.D.Faddeev, A remark on Schrödinger equation with a singular potential, Sov. Math. Dokl. 2, 372-375 (1961).

6 V.D.Koshmanenko, An operator representation for nonclosable quadratic forms and the scattering problem, Soviet Math. Dokl. 20, 294-297 (1979)

7 V.D.Koshmanenko, A classification of singular perturbation of selfadjoint operators, preprint 82, 34, Institute of Math., Kiev, 1982 (in Russian).

8 B.Simon, A cannonical decomposition for quadratic forms with applications to monotone convergence theorems, J. Funct. Anal. 28, 377-385 (1978).

9 M.G.Krein, The theory of self-adjoint extensions of semibonded Hermitian transformations and its applications. I, Rec. Math. (Math. Sb.), 20 (62), 431-495 (1947) (in Russian).

10 M.Sh.Birman, On the self-adjoint extensions of positive definite operators, Math. Sb. 38, 431-450 (1956) (in Russian).

11 W.G.Faris, Self-Adjoint Operators, Lecture Notes in Math. 433, 1975.

12 A.Alonso, B.Simon, The Birman-Krein-Vishik theory of self-adjoint extensions of semibounded operators, J. Oper. Th. 4, 251-270 (1980)

13 V.D.Koshmanenko, On the uniqueness on a singularly perturbed operator, to appear in Acad. Sci. USSR Dokl. (1988).

QUANTUM FIELD THEORY

Covariant markovian random fields in four space-time dimensions with nonlinear electromagnetic interaction

by

Sergio Albeverio*,**,***, Raphael Høegh-Krohn♯,†,
Koichiro Iwata*,♯♯

Abstract

We construct covariant random vector fields over 4-dimensional space-time as solutions of a system of first order coupled stochastic partial differential equations, best interpreted as equations for quaternionic valued random fields. The fields are covariant under the proper Euclidean transformations. We give necessary and sufficient conditions in terms of a given source of the infinitely divisible type, for the fields to be covariant also under reflections. In the case of a Gaussian white noise source the fields are Euclidean free electromagnetic potential fields and have the global Markov property. The fields with Poisson white noise source can be used as approximation of the Gaussian fields, with better support properties.

* **Fakultät für Mathematik, Ruhr-Universität, D 4630 Bochum 1 (FRG); SFB 237 Bochum-Essen-Düsseldorf**
** **BiBos Research Centre**
*** **CERFIM Research Centre, Locarno (CH)**
♯ **Matematisk Institutt, Universitetet i Oslo, Oslo (N)**
♯♯ **DAAD-Fellowship**
† **Deceased 24. January 1988**

To appear in Proc. Dubna Conference 1987, Edts. P. Exner, P. Seba, Lect. Notes Phys., Springer (1988)

1. Introduction

Local relativistic quantum field theory was invented more than sixty years ago to provide a synthesis of quantum mechanics and the (special) theory of relativity. In recent years models of local interacting relativistic quantum fields of scalar, vector or gauge type have been constructed in space-times of dimension less than 4, see e.g. [AHK1], [AFHKL], [DST], [GJ], [JLM].

In the case of 4-dimensional space times only partial results are known, see e.g. [DST], [AFHKL].
In the present paper we exhamine the possibility of constructing a four space-time dimensional theory describing quantum fields of the electromagnetic type, with a formal action which is not necessarily of the canonical type "kinetic energy minus potential term", but rather kinetic energy minus a term involving a nonlinear function of suitable linear combinations of derivatives of the field.
There is some relation of such models with those of non linear electromagnetic field theories, like Born-Infeld theory [BI]. Such nonlinear electromagnetic field theories had been introduced as approximations to Maxwell fields and our models can also be looked upon in the same spirit (and we prove indeed a result in this sense). Let us also remark that very recently the interest of Born-Infeld's action has been reactivated by the discovery that it describes heuristically the full effective self-interaction of vector fields in the Abelian limit in open bosonic strings (and superstrings), see e.g. [FT], [CLNY], [CF].
Our models exploit in an essential way the 4-dimensionality of the physical space-time, which permits to identify it, as a vector space, with the space $\mathbb{H}$ of quaternions[1)].
The fields are given as solutions of a system of coupled stochastic first order partial differential equations, having a natural formulation in terms of quaternions. The possibility of writing such equation relies on the isomorphism $SO(4) \cong (SU(2) \times SU(2))/\mathbb{Z}_2$. The Euclidean vector generalized random fields $\{A_r(x)\ ,\ x \in \mathbb{R}^4 \cong \mathbb{H}\ ,\ r = 0,1,2,3\}$, identified with quaternion fields $A(x)$, satisfy stochastic partial differential equations of the form $\partial A(x) = F(x)$, with $F(x)$ a quaternionic-valued infinitely divisible field (see e.g. [Kl], [Ku], [Su]) with suitable transformation properties under the proper Euclidean group $SO(4) \wedge \mathbb{R}^4$, ∂ being the basic 1-order quaternionic differential operator with unit coefficients[2)].
We discuss the transformation properties of A under reflections as well as Markovian properties of the fields. In the case of F being Gaussian white noise A is the free electromagnetic Euclidean potential field. We exhibit a way to approximate the latter field by fields A_p defined by taking F to be a Poisson type white noise.
We also point out that the fields A can be obtained as continuum limits of corresponding lattice fields, which makes appear their action as being heuristically given by
$\int f\left(|\mathrm{div}\, A|\, , |\vec{E} - \vec{B}|\right) dx$, with $A(x) = \left(A_0(x)\, , \vec{A}(x)\right)$, $x = (x_0, \vec{x}) \in \mathbb{R} \times \mathbb{R}^3$,
$\vec{E} \equiv \frac{\partial}{\partial x_0} \vec{A} - \mathrm{grad}_{\vec{x}} A_0\ ,\ \vec{B} \equiv \mathrm{rot}_{\vec{x}}\, A$, for suitable real valued functions f on $\mathbb{R}$.
Let us also remark that the present work is connected with previous work (see e.g. [AHKH1-3], [AHK 6], [AHKHK], [Ka] and references therein) in which Markov and quantum fields associated to 1-codimensional hypersurfaces, instead of points, in $\mathbb{R}^d$ were constructed. For $d = 2$ such "cosurface fields" can be identified, on closed contours, with

quantum gauge fields; for $d = 4$ they include free electromagnetic fields and more generally 3-forms with values in the Lie algebra of compact semisimple Lie groups, providing (by duality) a natural extension of electromagnetic fields to "coloured fields" (this relies on the realization of $\mathbb{R}^4$ and the Lie algebra $u(2)$ of $U(2)$ as the space of quaternions [AHK3]). The constructed cosurface can also be connected to vector fields, using again the 4-dimensionality of space-time, and these fields satisfy the stochastic partial differential equation discussed ([AHK2]).

We finally remark that the present paper extends the work of [AHK4] and makes precise the point first overlooked in [AHK4a] (but shortly remarked in [AHK4b]) that A is not time reflection invariant in the non Gaussian case.

2. A covariant quaternionic partial differential equation

We shall consider a covariant partial differential equation over $\mathbb{R}^4$. This type of equations can only be considered over $\mathbb{R}^1$, $\mathbb{R}^2$ and $\mathbb{R}^4$ and their existence is tied to that of the associative fields of real, complex resp. quaternionic numbers. In this paper we shall consider the physical situation with underlying space $\mathbb{R}^4$, and the equation is best formulated using quaternions, as in [AHK2-4]. Let $\mathbb{H}$ be the field of quaternionic numbers and $\{1, i, j, k\}$ be its canonical basis.

As a real vector space $\mathbb{H}$ is isomorphic to $\mathbb{R}^4$ by

$$\mathbb{H} \ni x_0 1 + x_1 i + x_2 j + x_3 k \longleftrightarrow (x_0, x_1, x_2, x_3) \in \mathbb{R}^4 .$$

We regard $\mathbb{R}$ as being inbedded in $\mathbb{H}$ by identifying $t \in \mathbb{R}$ with $t1 \in \mathbb{H}$, then $\mathbb{H}$ forms a real associative algebra with the identity 1 under the multiplication rules : $i^2 = j^2 = k^2 = -1$ and $ij = -ji = k$.

There is a distinct automorphism of $\mathbb{H}$ called the conjugation :

$$x = x_0 + x_1 i + x_2 j + x_3 k \longrightarrow \bar{x} = x_0 - x_1 i - x_2 j - x_3 k .$$

As in the case of $\mathbb{C}$ we write

$$Re\, x := \frac{1}{2}(x + \bar{x}) = x_0$$
$$Im\, x := \frac{1}{2}(x - \bar{x}) = x_1 i + x_2 j + x_3 k .$$

Later we also use the notation $\vec{x}$ for $Im\, x$. We see that the square root of the nonnegative quantity $x\bar{x} = \bar{x}x$ is equal to $|x|$, the $\mathbb{R}^4$-norm of x, under the above mentioned isomorphism $\mathbb{H} \cong \mathbb{R}^4$, and moreover

$$x \cdot y := \frac{1}{4}\left(|x+y|^2 - |x-y|^2\right) = Re\, x\bar{y} = Re\, \bar{x}y .$$

$Sp(1) := \{a \in \mathbb{H} \ ; \ |a| = 1\}$ is a subgroup of the multiplicative group $\mathbb{H}^{\times} := \mathbb{H} \setminus \{0\}$ and it is isomorphic to $SU(2)$. By $\mathbb{H} \ni x \longmapsto axb^{-1} \in \mathbb{H}$ for $a, b \in Sp(1)$ we have a surjective homomorphism $Sp(1) \times Sp(1) \longrightarrow SO(4)$, whose kernel is $\{(1,1) \ , \ (-1,-1)\} \cong \mathbb{Z}_2$, and hence $[Sp(1) \times Sp(1)]/\mathbb{Z}_2 \cong SO(4)$.

We consider the following two distinct $Sp(1) \times Sp(1)$ actions on $\mathbb{R}^4$-valued functions on $\mathbb{R}^4$: identifying $\mathbb{R}^4$ with $\mathbb{H}$, the first one is given by

$$A(x) \longrightarrow a\, A\left(a^{-1}(x-y)\, b\right) b^{-1} \qquad x, y \in \mathbb{R}^4 \ , \ (a,b) \in Sp(1) \times Sp(1) \qquad (i)$$

and A obeying this rules is called a <u>covariant 4-vector field</u>.

The second one is given by

$$A(x) \longrightarrow b\, A\left(a^{-1}(x-y)\, b\right) b^{-1} \qquad x, y \in \mathbb{R}^4 \ , \ (a,b) \in Sp(1) \times Sp(1) \qquad (ii)$$

and A obeying this rule is called a covariant scalar 3-vector field.
We define a bilinear form by

$$\langle \xi, A\rangle := \int_{\mathbb{R}^4} \xi(x)\cdot A(x)\,dx = Re\int_{\mathbb{R}^4} \xi(x)\cdot \overline{A(x)}\,dx \qquad \xi \in C^\infty{}_0(\mathbb{R}^4)$$

and extend this as the distributional pairing in the natural way. Note that $\langle\cdot,\cdot\rangle$ is invariant under $Sp(1)\times Sp(1)$ actions (i) and (ii).
Let

$$\partial := \frac{\partial}{\partial x_0} - i\frac{\partial}{\partial x_1} - j\frac{\partial}{\partial x_2} - k\frac{\partial}{\partial x_3} \text{ and } \bar\partial := \frac{\partial}{\partial x_0} + i\frac{\partial}{\partial x_1} + j\frac{\partial}{\partial x_2} + k\frac{\partial}{\partial x_3},$$

then $\partial\bar\partial = \bar\partial\partial = \Delta$, the Laplacian in $\mathbb{R}^4$. Consider two variables $x, x' \in \mathbb{R}^4$ related by $x' = a^{-1}xb$ for some $(a,b) \in Sp(1)\times Sp(1)$ and define ∂' and $\overline{\partial'}$ in the same way as ∂ and $\bar\partial$. Then it is easily seen that $\overline{\partial'} = a^{-1}\bar\partial b$ and $\partial' = b^{-1}\partial a$. Therefore, if A is a covariant 4-vector field, then $F = \partial A$ is a covariant scalar 3-vector field. This is well understood, if we introduce a 1-form $\sigma := \sum_{i=0}^{3} A_i\, dx_i$, the orientation adapted to $\{1, i, j, k\}$ and the associated Hodge duals. In fact, identifying anti-self dual 2-forms with 3-vector fields, we have $({}^*d^*\sigma\ ,\ d\sigma - {}^*d\sigma) = (F_0\ ,\ \vec{F})$.
We note that the equation $\partial A = F$ is not covariant under reflections, since $\vec{F}$ corresponds to an anti-self dual 2-form.
We denote by g Green′ s function to $-\Delta$, i.e., $g(x) = \dfrac{1}{2\pi^2|x|^2}$ and set

$$S(x) := -\bar\partial\, g(x) = \frac{x}{\pi^2|x|^4} \quad , \quad \bar S(x) := \overline{S(x)}\,,$$

then we see that

$$\partial S(x) = -\partial\bar\partial\, g(x) = -\Delta\, g(x) = \delta(x) \quad , \quad \bar\partial \bar S(x) = \delta(x)\,,$$

where δ is the Dirac distribution. In order to give a precise meaning to the inverse of ∂ (resp $\bar\partial$) we introduce the following space

$$\mathcal{I} := \left\{\varphi \in C^\infty(\mathbb{R}^4, \mathbb{H})\,;\ \lim_{|x|\to+\infty} \varphi(x) = 0\,,\ \bar\partial\varphi \in \mathcal{S}\right\}$$

($\mathcal{S} \equiv \mathcal{S}(\mathbb{R}^4, \mathbb{H})$ is the Schwartz test space of rapidly decreasing test functions).
It is easily seen that $\mathcal{I} \ni \varphi \longrightarrow \bar\partial\varphi \in \mathcal{S}$ is bijective and the inverse map is given by $\mathcal{S} \ni \xi \longrightarrow \bar S * \xi \in \mathcal{I}$, where

$$\bar S * \xi(x) = \int_{\mathbb{R}^4} \bar S(x-y)\,\xi(y)\,dy\ .$$

Using this isomorphism we introduce a locally convex topology on $\mathcal{I}$. Note that the injection $\iota : \mathcal{S} \hookrightarrow \mathcal{I}$ is not dense and hence $\iota^* : \mathcal{I}' \longrightarrow \mathcal{S}'$ is not injective, since $\{\bar\partial\varphi\,;\ \varphi \in \mathcal{S}\}$ is not dense in $\mathcal{S}$.
We have the

Theorem 1 : Let A be a covariant 4-vector field and F be a covariant scalar 3-vector field. Then the elliptic 1-order partial differential equation

$$\partial A(x) = F(x)$$

is covariant. If F belongs to $\mathcal{I}'$, $\partial A = F$ has a unique solution in $\mathcal{S}'$ given by $A = S * F$.

We interpret A as a classical electromagnetic Euclidean potential (in the Feynman gauge) and $E_i := \partial_0 A_i - \partial_i A_0$, $i = 1,2,3$, resp. $B_i = \partial_j A_k - \partial_k A_j$ (i,j,k) cyclic permutation of $(1,2,3)$ as electro resp. magnetic fields. $\vec{F}$ corresponds to $E_i - B_i$.

3. Random fields as solutions of a quaternionic partial differential equation with random source

We shall now consider the equation $\partial A = F$ in sect. 2, in the case where F is a generalized random field over $\mathbb{R}^4$ with values in $\mathbb{H}$. We assume that $\{F(x)\}$ and $\{bF(a^{-1}(x-y)b)b^{-1}\}$ have the same finite dimensional distributions for all $((a,b),y) \in Sp(1) \times Sp(1) \times \mathbb{R}^4$ and we call such F an invariant scalar 3-vector (generalized) random field. From the result in sect. 2 we see that the $\mathbb{H}$-valued (generalized) random field A related to F by the equation $\partial A = F$ is invariant, in the sense of law, under proper Euclidean transformations. We shall call such A an invariant 4-vector (generalized) random field (or also, for short, as in the title, a covariant random field). We have:

Theorem 2 : If F is an invariant scalar 3-vector generalized random field realized as a $\mathcal{I}'$-valued random variable, then $\partial A = F$ has a unique solution $A = S * F$ realized as an $\mathcal{S}'$-valued random variable. A is an invariant 4-vector random field. ∎

In what follows we further assume that F is independent at every point, i.e., if we restrict its characteristic functional to $\mathcal{S}$, then taking translation invariance into account we have

$$C_F(\varphi) := E\left[e^{\sqrt{-1}\langle\varphi,F\rangle}\right] = \exp\left(-\int_{\mathbb{R}^4} \psi(\varphi(x))\, dx\right) \quad \varphi \in \mathcal{S}$$

with ψ a continuous negative definite function on $\mathbb{H}$. Because of its $Sp(1)$ adjoint invariance, $\psi(b\lambda b^{-1}) = \psi(\lambda)$, ψ has the following Lévy-Khinchine representation :

$$\psi(\lambda) = -\sqrt{-1}\,\lambda_0\beta + \frac{\sigma_0}{2}\lambda_0^2 + \frac{\sigma}{2}|\vec{\lambda}|^2 + \int \left(1 + \sqrt{-1}\,\lambda\cdot\alpha\, \chi_{(0,1)}(|\alpha|) - e^{\sqrt{-1}\,\lambda\cdot\alpha}\right)\nu(d\alpha), \quad \lambda \in \mathbb{H}$$

with $Sp(1)$ adjoint invariant Lévy measure ν $\left(\nu(b\, d\alpha\, b^{-1}) = \nu(d\alpha)\right)$ and $\beta \in \mathbb{R}$, $\sigma_0, \sigma \geq 0$. We call ψ resp. $(\beta, \sigma_0, \sigma, \nu)$ the Lévy characteristics of F. If it is possible to extend the domain of $C_F(\cdot)$ to $\mathcal{I}$, then F is realizable as a $\mathcal{I}'$-valued random variable. To this end we assume that $\psi(\lambda) = O\left(|\lambda|^{\frac{4}{3}+\varepsilon}\right)$ as $\lambda \to 0$ for some $\varepsilon > 0$. Indeed under this assumption the characteristic function $C_F(\cdot)$ defined on $\mathcal{I}$ is uniquely determined by $\exp\left(-\int_{\mathbb{R}^4} \psi(\varphi(x))\, dx\right)$ $, \varphi \in \mathcal{S}$, and consequently the $\mathcal{I}'$-valued random variable F is uniquely characterized by the $\mathcal{S}'$-valued random variable $\iota^* \circ F$. For we see from Sobolevś inequality that

$$\|\bar{S} * \xi(x)\|_{L^p} \leq \left\{\int\left|\int \frac{|\xi(y)|}{\pi^2|x-y|^3}\, dy\right|^p dx\right\}^{\frac{1}{p}} \leq C_p\|\xi\|_{L^q}\,, \quad p > \frac{4}{3}\,, \ \frac{1}{q} = \frac{1}{p} + \frac{1}{4}\,.$$

Theorem 3 : Let F be a translation invariant $\mathbb{H}$-valued generalized random field over $\mathbb{R}^4$ independent at every point with Lévy characteristic ψ. Then F has the properties as in Theorem 2 and A solving $\partial A = F$ has a distribution given by :

$$C_A(\xi) := E\left[e^{\sqrt{-1}\langle \xi, A\rangle}\right] = \exp\left(-\int_{\mathbb{R}^4} \psi(-\bar{S} * \xi(x))\, dx\right) \quad , \ \xi \in \mathcal{S}\,.$$

In particular A is an invariant 4-vector generalized random field. If the Lévy measure ν has the p-th moment, i.e. $\int |\alpha|^p \nu(dp) < \infty$ for $p = 2, 3, \ldots$, then A also has the p-th moment

$$E\left[\langle \xi_1, A\rangle \ldots \langle \xi_p, A\rangle\right] \quad , \ \xi_1, \ldots, \xi_p \in \mathcal{S}\,,$$

as a continuous linear functional on $\mathcal{S}^{\otimes p}$.

We call such F with the properties in Theorem 3 an <u>invariant scalar 3-vector generalized random field of the infinitely divisible type</u>.

Concerning reflection invariance of A, the situation is utterly changed according to whether F is Gaussian distributed or not. By a reflection we mean the following $\mathbb{Z}_2$-action on covariant 4-vector fields :

$$\rho : \quad A(x) \longrightarrow -\overline{A(-\bar{x})} \quad , \quad x \in \mathbb{R}^4\,.$$

If A is ρ-invariant as well, then A is invariant under full Euclidean transformations. Before going into general cases, we first note that A is invariant under the reflection ρ when $\psi(\lambda)$ depends only on $\lambda_0 = Re\ \lambda$, i.e., $\sigma = 0$ and ν is supported by $\mathbb{R}\backslash\{0\} = \{\alpha \in \mathbb{H}^\times\,;\, Im\ \alpha = 0\}$. Indeed since the operators $g * \cdot$ and div $\cdot$ commute with ρ, we have

$$\begin{aligned} -Re\ \bar{S} * \rho\xi(x) &= Re\ \partial g * \rho\xi(x) = g * \ Re\ \partial(\rho\xi(x)) \\ &= g * \ \mathrm{div}\ \rho\xi(x) = -Re\ \bar{S} * \xi(-\bar{x}) \end{aligned}$$

for $x \in \mathbb{R}^4$, $\xi \in \mathcal{S}$, and therefore

$$\begin{aligned} C_A(\rho\xi) &= \exp\left\{-\int \psi\left(-\bar{S} * \rho\xi(x)\right) dx\right\} \\ &= \exp\left\{-\int \psi\left(-\bar{S} * \xi(-\bar{x})\right) dx\right\} = C_A(\xi) \end{aligned}$$

We shall now discuss the case of a pure Poisson source, i.e., $\sigma_0 = \sigma = 0$ and $\nu \neq 0$. We assume that ν has a compact support in $\mathbb{H}^\times$ and $\beta = 0$ for simplicity. Let $\{(\alpha_i, x_i)\}_{i=1}^{\infty}$ be the Poisson point process on $\mathbb{H}^\times \times \mathbb{R}^4$ with Lévy measure $\nu(d\alpha) \otimes dx$, then F is realizable as an $\mathbb{H}$-valued random measure over $\mathbb{R}^4$ by using $\{(\alpha_i, x_i)\}_{i=1}^{\infty}$:

$$F(x) = \sum_{i=1}^{\infty} \alpha_i \, \delta_{\{x_i\}}(x)$$

and therefore A solving $\partial A = F$ one has the following representation :

$$A(x) = S * F(x) = \sum_{i=1}^{\infty} \frac{x - x_i}{\pi^2 |x - x_i|^4} \alpha_i \quad .$$

We now see what happens if we perform the reflection ρ. Because of the invariance of the Lévy measure $\nu(d\alpha) \otimes dx$ the law of $\{(\bar{\alpha}_i, -\bar{x}_i)\}_{i=1}^{\infty}$ is equal to that of $\{(\alpha_i, x_i)\}_{i=1}^{\infty}$, so that we have

$$\rho A(x) \equiv -\overline{A(-\bar{x})} = -\sum_{i=1}^{\infty} \bar{\alpha}_i \, \frac{\overline{(-\bar{x} - x_i)}}{\pi^2 |-\bar{x} - x_i|^4} = \sum_{i=1}^{\infty} \bar{\alpha}_i \, \frac{x - (-\bar{x}_i)}{\pi^2 |x - (-\bar{x}_i)|^4} \stackrel{\mathrm{d}}{=} \sum_{i=1}^{\infty} \alpha_i \, \frac{x - x_i}{\pi^2 |x - x_i|^4} \, ,$$

where $\stackrel{\mathrm{d}}{=}$ stands for the law equivalence. Suppose that $\rho A \stackrel{\mathrm{d}}{=} A$, then $\partial \rho A \stackrel{\mathrm{d}}{=} \partial A = F$. However this does not hold unless $Im\ \alpha_i = 0 \ \forall i$ a.s., since it follows from the definition of S that

$$\begin{aligned}
\partial \rho A(x) &\stackrel{\mathrm{d}}{=} \partial \left(\sum_{i=1}^{\infty} \alpha_i \, \frac{x - x_i}{\pi^2 |x - x_i|^4} \right) \\
&= \partial \sum_{i=1}^{\infty} \left(\alpha_i \frac{x - x_i}{\pi^2 |x - x_i|^4} - \frac{x - x_i}{\pi^2 |x - x_i|^4} \alpha_i \right) + \sum_{i=1}^{\infty} \alpha_i \delta_{\{x_i\}}(x)
\end{aligned}$$

and hence there are subsets of $\mathcal{S}'$ which have zero measure for $\partial(\rho A)$ and measure 1 for F. Hence A in the pure Poisson case is not ρ-invariant, unless $Im\ \alpha_i = 0 \ \forall i$ a.s. .

Let us look as a contrast to the case F Gaussian, i.e. $\nu = 0$, and see how one recovers the reflection invariance. We have

$$\begin{aligned}
\int_{\mathbb{R}^4} |\bar{S} * \xi(x)|^2 \, dx &= \langle \partial g * \xi, \partial g * \xi \rangle = -\langle g * \xi, \bar{\partial} \partial g * \xi \rangle \\
&= \langle g * \xi, \xi \rangle \\
&= \int g(x - y) \, \xi(x) \cdot \xi(y) \, dx dy
\end{aligned}$$

and thus we have

$$\int_{\mathbb{R}^4} |\bar{S} * \rho\xi(x)|^2 dx = \int_{\mathbb{R}^4} |\bar{S} * \xi(x)|^2 dx \quad , \xi \in \mathcal{S} \, .$$

Combining this with the fact $Re\ \bar{S} * \rho\xi(x) = Re\ \bar{S} * \xi(-\bar{x})$ we get

$$\int_{\mathbb{R}^4} |Im\ \bar{S} * \rho\xi(x)|^2 dx = \int_{\mathbb{R}^4} |Im\ \bar{S} * \xi(x)|^2 dx \quad , \xi \in \mathcal{S} \, .$$

This implies $C_A(\rho\xi) = C_A(\xi)$, $\xi \in \mathcal{S}$, as far as ν vanishes. This fact is a striking contrast to the case of a pure Poisson source. We summarize these results in the following

Theorem 4 : Let F be an invariant scalar 3-vector generalized random field of the infinitely divisible type. Then A solving $\partial A = F$ is ρ-invariant iff the Lévy measure ν associated with F is supported by $\mathbb{R}\backslash\{0\}$.

From the point of view of the Euclidean field theory, A for $\beta = 0$, $\sigma_0 = \sigma = 1$, $\nu = 0$ corresponds to the free electromagnetic potential field in the Feynman gauge and A for $\beta = 0$, $\sigma_0 = 0$, $\sigma = 1$, $\nu = 0$ corresponds to that in the Coulomb gauge.

4. Some further properties of the constructed random fields

It is well known that the passage from Euclidean fields to relativistic fields is possible in general situations where the Osterwalder-Schrader (reflection) positivity (see e.g. [GJ]) holds under time reversal ρ. In the following we shall first see how we can construct from A an Osterwalder-Schrader ($\equiv$ O.S.) positive field by taking into account the gauge invariance of the underlying equation $\partial A = F$ (if A is a solution, then $A + \bar{\partial}\chi$ with χ harmonic, i.e. $\Delta\chi = 0$, also solves $\partial A = F$) in the Gaussian case, $\nu = 0$ in the notation of section 3. Let $\mathcal{S}_{df}$ be the subspace of $\mathcal{S}$ consisting of all ξ with $Re\ \partial\xi = \text{div}\ \xi = 0$. We note that $\mathcal{S}_{df}$ is a Euclidean invariant test function space and $\{\langle\xi, A\rangle\,;\ \xi \in \mathcal{S}_{df}\}$ is a family of gauge invariant random variables, which we shall call the Euclidean transversal field with the gauge potential A. Since $Re\ \bar{S} * \xi = Re\ g * \partial\xi = 0$ for $\xi \in \mathcal{S}_{df}$, the covariance functional of the transversal field is equal to

$$\sigma \int |\bar{S} * \xi(x)|^2 dx = \sigma \int g(x-y)\xi(x) \cdot \xi(y)\, dx dy \equiv \sigma(\xi, \xi)\ ,$$

whatever the parameter σ_0 is. Suppose that $\xi \in \mathcal{S}_{df}$ has its support in $\mathbb{R}_+ \times \mathbb{R}^3$. By using the partial Fourier transformation, we have

$$(\xi, \rho\xi) = \int_{\mathbb{R}^3} \int_0^\infty e^{-|\vec{k}|t} \widehat{\xi(t,\cdot)}(\vec{k}) dt \cdot \int_{-\infty}^0 e^{|\vec{k}|s} \widehat{\rho\xi(s,\cdot)}(-\vec{k}) ds \frac{d\vec{k}}{2|\vec{k}|}\ .$$

Next we apply the integration by parts formula to the dt-integral, then, since div $\xi = 0$, it follows that

$$\int_0^\infty e^{-|\vec{k}|t} \widehat{\xi_0(t,\cdot)}(\vec{k}) dt = -\sqrt{-1} \int_0^\infty e^{-|\vec{k}|t} |\vec{k}|^{-1} \vec{k} \cdot \widehat{\vec{\xi}(t,\cdot)}(\vec{k}) dt$$

and similary

$$\int_{-\infty}^0 e^{|\vec{k}|s} \widehat{(\rho\xi)_0(s,\cdot)}(-\vec{k}) ds = -\sqrt{-1} \int_0^\infty e^{-|\vec{k}|s} |\vec{k}|^{-1} \vec{k} \cdot \widehat{\vec{\xi}(s,\cdot)}(-\vec{k}) ds\ .$$

Hence we obtain the positivity

$$(\xi, \rho\xi) = \int_{\mathbb{R}^3} \left\{ \left| \int_0^\infty e^{-|\vec{k}|t} \widehat{\vec{\xi}(t,\cdot)}(\vec{k}) dt \right|^2 - \left| \int_0^\infty e^{-|\vec{k}|t} |\vec{k}|^{-1} \vec{k} \cdot \widehat{\vec{\xi}(t,\cdot)}(\vec{k}) dt \right|^2 \right\} \frac{d\vec{k}}{2|\vec{k}|} \geq 0\ .$$

This implies the O.S.-positivity of the transversal field : $\forall z_i \in \mathbb{C}$, $i = 1, \ldots, n$

$$\sum_{i,j=1}^{n} z_i \bar{z}_j E\left[e^{\sqrt{-1}\langle \xi_i, A\rangle}\, e^{-\sqrt{-1}\langle \xi_j, \rho A\rangle}\right] \geq 0 \quad , \quad \xi_j \in \mathcal{S}_{df}\, , \, \mathrm{supp}\, [\xi_j] \subset \mathbb{R}_+ \times \mathbb{R}^3 \, .$$

As usual as we obtain the physical Hilbert space $\mathcal{H}$ spanned by $\left\{e^{\sqrt{-1}\langle \xi, A\rangle} ; \, \xi \in \mathcal{S}_{df}\, , \, \mathrm{supp}\, [\xi] \subset \mathbb{R}_+ \times \mathbb{R}^3\right\}$ with the inner product naturally introduced by the O.S.-positive condition and the symmetric contraction semigroup acting on $\mathcal{H}$, which is determined by

$$E\left[e^{\sqrt{-1}\langle \xi, \tau_{-t} A\rangle}\, e^{-\sqrt{-1}\langle \xi, \rho A\rangle}\right] \, ,$$

where $\{\tau_t\}$ is the shift along the x_0-axis : $\tau_t A(x) = A(x_0 - t, \vec{x})$.
The negative H of the generator of the above semigroup is the physical energy operator. Using this operator we can construct relativistic potential fields as operator-valued distributions (with test functions in $\mathcal{S}_{df}$). These fields can be identified with the electromagnetic free potential fields.

Remark 1 : It is possible to show that above Euclidean transversal electromagnetic potential fields $\{\langle \xi, A\rangle\}$ have the Markov property with respect to arbitrary open subsets of $\mathbb{R}^4$, in the sense that, extending $\langle \xi, A\rangle$ to all $\xi \in S'$ with $(\xi, \xi) < \infty$ and denoting by $\sum_\Lambda \equiv \sigma(\langle \xi, A\rangle; \xi \in S', (\xi,\xi) < \infty, \mathrm{div}\, \xi = 0, \mathrm{supp}[\xi] \subset \Lambda) \vee \mathcal{N}$ the σ-algebra generated by the fields in the Borel region Λ, and the zero measure sets $\mathcal{N}$ (with respect to the measure associated with A), then for any open $D \subset \mathbb{R}^4$, $\sum_{\bar{D}}$ is conditionally independent of $\sum_{D^c}$ given $\sum_{\partial D}$, where $\bar{D}$ is the closure of D, $D^c \equiv \mathbb{R}^4 - D$, ∂D is the boundary of D.
This is proven using the Fock space or Wiener-chaos decomposition of $\mathcal{H}$. For some related discussions see e.g. [Lö] and references therein.
It is also known that the Markov property holds even for $\{\langle \xi, A\rangle\}$, with ξ not restricted to be in $\mathcal{S}_{df}$ ("non transversal fields "), provided one takes the Feynman gauge $\beta = 0$, $\sigma_0 = \sigma = 1$, $\nu = 0$. In fact this holds also for any Gaussian field defined by

$$E\left(e^{\sqrt{-1}\langle \xi, A\rangle}\right) = e^{-\frac{1}{2}(\xi,\xi)_c}\, , \, \xi \in \mathcal{S}\, , \, 0 \leq c < 1 \, ,$$

with

$$(\xi, \xi)_c \equiv \int_{\mathbb{R}^4} \left[|\hat{\xi}(k)|^2 - c\frac{|k \cdot \hat{\xi}(k)|^2}{|k|^2}\right] |k|^{-2}\, dk$$

(with $\hat{\xi}$ the Fourier transform of ξ and c the ratio σ_o/σ in the notation of section 3). $c = 1$ corresponds to Coulomb gauge, where one only has the Markov property when restricting ξ to be in $\mathcal{S}_{df}$.
That even in the non purely Gaussian case $\nu \neq 0$ one should still have Markovian properties is suggested by the fact that ∂ is a first order partial differential operator. However this is not yet fully mathematically settled. One difficulty is due to the bad spectral properties of ∂^{-1} ($\partial A = F$ being a "zero mass" equation).
In related positive mass equations it is possible to prove the 0-Markov property in the sense of Kusuoka [K], see [I]. Let us also remark that Surgailis has discussed related problems in the case where $\mathbb{R}^4$ is replaced by $\mathbb{R}^2$, see [Su2].

Remark 2 : As remarked in [AHK3], it is possible to associate to the quaternionic valued field A in the general case, a component wise 3-form $\omega = (\omega_\mu\,,\, \mu = 0,1,2,3)$.
In fact let $a_0 \equiv A\,,\, a_1 \equiv -iA\,,\, a_2 = -jA\,,\, a_3 = -kA$. Then $a \equiv \sum_\mu a_\mu dx_\mu$ is also a quaternionic valued 1-form. We have $\sum_{\mu=0}^{3} \partial_\mu\, a_\mu = \partial A = F$.
Let ω be the Hodge dual of a, then $d\omega = F$, in the sense that $d\omega_\mu = F_\mu\,,\, \mu = 0,1,2,3$, where ω_μ is looked upon as a 3-form over $\mathbb{R}^4$ and F_μ is looked upon as a 4-form over $\mathbb{R}^4$. $d\omega = F$ can be written as $\omega(\partial B) = F(B)$, for any measurable $B \subset \mathbb{R}^4$, where by definition $\omega(\partial B) = \int_B d\omega$ (in analogy with the corresponding formulae which hold when ω and B are smooth). ω is then a Markov Euclidean invariant cosurface in the sense of [AHKH1], a stochastic integral in the sense of [AHKH2]. In [AHK3] the relation $d\omega = F$ is extended to the case where the ω_μ are 3-forms with values in a Lie algebra g containing that of $U(2)$. ω is then a g-valued Markov cosurface.

Remark 3 : All considerations of this section, with invariance properties suitable reinterpreted, can also be made for the case where the region $\mathbb{R}^4$ on which the fields are defined is replaced by an open domain B with boundary ∂B. Let in fact $(F\,,\, P_B)$ be the generalized random field defined by
$E_{P_B}(e^{\sqrt{-1}\langle \xi, F\rangle}) = \exp\left(-\int_B \psi\,(\xi(x))\,dx\right)$, with ψ as in Theor. 2. Let ∂_B be defined by closure in $L^2(dx)$ from ∂ on $C^\infty{}_0(B\,;\,\mathbb{R}^4)$. Let S_B be the fundamental solution to ∂_B. S_B has the same local behavior as S. The analogue of Theor. 2 holds then with S replaced by S_B, yielding a solution of the equation $\partial_B\, A = F$. A is rotation invariant if B is rotation invariant.
Let μ_B be the probability measure giving the distribution of the field A. (A, μ_B) is a locally Markov field in the sense of [AHK7], [Nc]. (A, μ_B) converges weakly as $B \uparrow \mathbb{R}^4$ to (A, μ), with (A, μ) given by Theor. 2.

Remark 4 : It is possible to discuss a "lattice approximation" of the field A constructed in Sect. 3.
Let $\delta > 0$, $\mathbb{Z}^4_\delta \equiv \{\delta n\,,\, n \in \mathbb{Z}^4\}$, $\Lambda_\delta \equiv \Lambda \cap \mathbb{Z}^4_\delta$ for any bounded subset Λ of $\mathbb{R}^4$.
Let $P_{\Lambda_\delta}(\cdot)$ be the probability measure on $\mathbb{H}^{\Lambda_\delta}$ given by

$$dP_{\Lambda_\delta}(F) \equiv (Z_\delta)^{-|\Lambda_\delta|} \exp\left(-W_\delta(F)\right) \prod_{x\in\Lambda_\delta} dF(x)\,,$$

with

$$W_\delta(F) \equiv \sum_{x\in\Lambda_\delta} \delta^4\, f_\delta\left(F_0(x)\,,\, |\vec{F}(x)|\right)\,,$$

with f_δ a positive function on $\mathbb{R}^4$ s.t. $f_\delta(\gamma) = f_\delta(|\gamma_0|, |\vec{\gamma}|)\ \forall \gamma \in \mathbb{R}^4$, $Z_\delta \equiv \int_{\mathbb{R}^4} e^{-\delta^4 f_\delta(\gamma)} d\gamma < \infty$ and

$$\lim_{\delta\downarrow 0} \delta^{-4} \left\{ \ln\left[\int_{\mathbb{R}^4} e^{-\sqrt{-1}\delta^4 \lambda\gamma}\, e^{-\delta^4 f_\delta(\gamma)}\, d\gamma / Z_\delta \right] \right\}$$

exists for all $\lambda \in \mathbb{R}^4$ and is a Lévy-Khinchine function having the same properties as the function ψ entering Theor. 2.

Remark 5 : An example is given by the convolution semigroup $\{P_t\}$, $t \geq 0$ of probability densities associated with ψ. Namely we choose $f_\delta(\gamma) \equiv -\delta^{-4} \log \{P_{\delta^4}(\delta^4\gamma)\delta^{16}\}$, then $\int e^{\sqrt{-1}\lambda\cdot\gamma} e^{-\delta^4 f_\delta(\gamma)} d\gamma = e^{-\delta^4\psi(\delta^{-4}\lambda)}$.
Let us define

$$d\mu_\delta(A) \equiv K_\delta e^{-W_\delta(\partial_\delta A)} \prod_{x\in\Lambda_\delta} dA(x)$$

with $A(x) : \Lambda_\delta \to \mathbb{H}$, K_δ a constant making $d\mu_\delta$ into a probability measure and ∂_δ a discrete version of ∂. It is possible to show that (A, μ_δ) converges weakly as $\delta \downarrow 0$ to the continuum limit (A, μ) described in Sect. 3.
Finally we remark that the field (A, μ) constructed from a Lévy characteristic $\psi = \psi_p$ of Poisson type can approximate the free electromagnetic Euclidean field arbitrary well.
In fact let us choose the Lévy characteristic ν_r of A to be in s.t., e.g. for r_0 , $r > 0$:

$$\nu_r(|\alpha_0|, |\vec{\alpha}|) = 3\ [\delta_{r_0}(\alpha_0) + \delta_{-r_0}(\alpha_0)]\ \rho(|\vec{\alpha}|)/(8\pi r^4),$$

with ρ the restriction of Lebesgue measure to $|\vec{\alpha}| = r$.
Then

$$-\psi(\lambda) = 3 \left[\cos\lambda_0 r_0 \int_{|\vec{\alpha}|=r} e^{\sqrt{-1}|\vec{\lambda}| r\cos\theta} r^2 (\sin\theta)\, d\theta d\varphi - 4\pi r^2\right] / (4\pi r^4) \underset{r_0, r\downarrow 0}{\longrightarrow} -\frac{\lambda^2}{2}\,.$$

Calling μ_r the probability measure given by ψ, we have that (μ_r, A) converges in this case for $r \to 0$ weakly to the free Euclidean electromagnetic potential field. This can be used to study interactions with matter, see [AIW].
Exploiting the support properties of A one can study local perturbations of the field (μ_p, A). Let v be a $\mathbb{R}$-valued Borel measurable function on $\mathbb{R}^4$ s.t., for $|\lambda| \to \infty$,

$$v(\lambda) = v(|\lambda|) = O(|\lambda|^\alpha)\ ,\ \alpha < \frac{4}{3}\,,$$

v bounded on compacts.
Let μ_B be as in Remark 3, with $\psi = \psi_p$. Then

$$\int_B v(|A(x)|)\, dx \in L^1(\mu_B),$$

for any $B \subset \mathbb{R}^4$ bounded measurable. Thus if in addition v is bounded from below, then

$$d\mu_B{}^v(A) \equiv Z_B^{-1}\, e^{-\int_B v(|A(x)|)\,dx}\, d\mu_B(A)\,,$$

with Z_B the normalizing constant, is a well defined probability measure. $(A, \mu_B{}^v)$ is locally Markov.
For v suitable, e.g. $v \geq 0$ one gets weak limits points as $B \nearrow \mathbb{R}^4$.
In this way we can create new locally Markov random fields, covariant under the proper Euclidean group.

Footnotes

[1] This is similar to the association of $\mathbb{R}^2$ with complex numbers. Our use of quaternions is different from one done in a large literature involving quaternionic (and octonionic) Hilbert spaces for the study of elementary particle models (see e.g. [A] and references therein). In fact our use is more similar to the one done in relation with classical electromagnetic fields, starting with Maxwell. Our approach has been partly announced in [AHK2–4]. On the basis of this announcement Osipov [O] has given an extension, renouncing of course associativity, to 8-space-time dimensions by using octonions.

[2] Euclidean (generalized) random fields as solutions of stochastic differential equations have been discussed before in [AHK2,3,4]. For lower space-time dimension or Gaussian fields (free fields) see [AHK1,5], [Ca], [GuL], [GuR], [Ha], [JLM], [Rö], [Su2] and references therein.

Acknowledgements

Raphael Høegh-Krohn reported on a previous version of this work in Dubna. On January 24, '88 Raphael suddenly died. In great sorrows we deeply mourn his departure and acknowledge our great indebtness to him. We also greatfully acknowledge great stimulation received from Prof. Z. Haba by his pointing out at an early stage the natural use of quaternionic calculus in electromagnetism. We are very grateful to Prof. S. Kusuoka for his patience and constructive criticism of a previous version of the paper. We also thank Professors B. Gawedzki, R. Gielerak and R. Streater for helpful criticism on previous versions of this work. The kind invitation of the first and second author to the Dubna Conference is gratefully acknowledged, as well as the DAAD support to the third author.

References

[A] Adler, S.: Quaternionic quantum field theory, Comm. Math. Phys. **104**, 611–65((1986)

[AFHKL] Albeverio, S.; Fenstad, J. E.; Høegh-Krohn, R.; Lindstrøm, T.: Nonstandard Method: in Stochastic Analysis and Mathematical Physics, Academic Press, Orlando (1986)

[AHK1] Albeverio, S.; Høegh-Krohn, R.: Diffusion fields, quantum fields and fields witl values in groups, in *Stochastic Analysis and Applications*, 1–98, Edt. M. Pinsky M. Dekker, New York (1984)

[AHK2] Albeverio, S.; Høegh-Krohn, R.: Euclidean Markov fields and relativistic quantun fields from stochastic partial differential equations in four dimensions, Phys. Letts **B177**, 175-179 (1986)

[AHK3] Albeverio, S.; Høegh-Krohn, R.: Quaternionic non abelian relativistic quantum field: in four space-time dimensions, Phys. Letter B 189, 329-336, (1987)

[AHK4] Albeverio, S.; Høegh-Krohn, R.: a) Construction of interacting local relativistic quan tum fields in four space-time dimensions, Phys. Letts. B, 200, 108-114 (1988) ; b) Err. Phys. Letts. B, Febr. (1988)

[AHK5] Albeverio, S.; Høegh-Krohn, R.: Quasi-invariant measures, symmetric diffusion pro cesses and quantum fields, Proc. Int. Coll. Math. Math. Quantum Field Theory CNRS **248** , 11–59 (1976)

[AHK6] Albeverio, S.; Høegh-Krohn, R.: Some recent interactions between mathematics an(physics in connection with generalized random fields, Proc. 1st World Congress o Bernoulli Society. Tashkent 1986, VNU-Press (1988)

[AHK7] Albeverio, S.; Høegh-Krohn, R.: Uniqueness and the global Markov property fo Euclidean fields. The case of trigonometric interactions, Comm. Math. Phys. 68 95–128 (1979)

[AHKH1] Albeverio, S.; Høegh-Krohn, R.; Holden, H.: Markov cosurfaces and gauge fields Acta Phys. Austr., Suppl. XXVI, 211–231 (1984)

[AHKH2] Albeverio, S.; Høegh-Krohn, R.; Holden, H.: Stochastic multiplicative measure, gen eralized Markov semigroups and group valued stochastic processes and fields, J. Funct Anal., 78, 154-184 (1987)

[AHKH3] Albeverio, S.; Høegh-Krohn, R.; Holden, H.: Random fields with values in Lie group and Higgs fields, in *Stochastic Processes in Classical and Quantum systems*, Proc Ascona, 1985, Edts. S. Albeverio, G. Casati, D. Merlini, Lect. Notes Phys. Springe Berlin (1986)

[AHKHK] Albeverio, S.; Høegh-Krohn, R.; Holden, H.; Kolsrud, T. : Representations an construction of multiplicative noise, Stockholm and BiBos-Preprint (1987), to appea in J. Funct. Anal. (1988)

[AIW] Albeverio, S.; Iwata, K.; Welz, B. : in preparation

[BI] Born, M.; Infeld, L.: Proc. Roy. Soc. A 144, **425** (1934)

[Ca] Carlen, E. : The stochastic mechanics of the scalar fields, pp. 40–60 in A. Trumar I.M. Davies (Edts.), *Stochastic Mechanics and Stochastic Processes*, Lect. Note Maths. 1325, Springer, Berlin (1988)

[CF] Cecotti, S.; Ferrara, S.: Supersymmetric Born-Infeld Lagrangians, Cern-Preprin Dec. 1986

[CLNY] Callan, C. G.; Lovelace, C; Nappi, C. R.; Yost, S. A.: Princeton Preprint PUPT-1027 (1986)

[DKK] Derrik, G. H.; Kay-Kong, W.: Particle motion and interaction in nonlinear field theories, J. Math. Phys. 9, 232–240 (1968)

[DST] Davies, I.M.; Simon, B.; Truman, A.: Edts. Proceedings IAMP Conf., Swansea (1988)

[FT] Fradkin, E. S.; Tseythin, A. A.: Phys. Lett. 163 B, **123** (1985)

[GJ] Glimm, J.; Jaffe, A.: *Quantum Physics*, 2n Ed., Springer, Berlin (1987)

[GuL] Guerra, F.; Loffredo, M. I.: Stochastic equations for the Maxwell Field, Lett. Nuovo Cim. **27**, 41–45 (1980)

[GuR] Guerra, F.; Ruggiero, P.: A new interpretation of the Euclidean-Markov field in the framework of physical Minkowski space-time, Phys. Rev. Letts. **31**, 1022-1025 (1973)

[Gr] Gross, L. : A Poincaré Lemma for connection forms, J. Funct. Anal. 63, 1 (1985)

[Ha] Haba, Z.: Stochastic equations for some Euclidean fields, in *Stochastic Processes in classical and Quantum System*, Proc. Ascona, Edts. S. Albeverio, G. Casati, D. Merlini, Lect. Notes Phys. **262**, 315–328, Springer, Berlin (1986)

[I] Iwata, K., work in preparation

[JLM] Jona-Lasinio, G.; Mitter, P. K.: On the stochastic quantization of field theory, Comm. Math. Phys. **101**, 409–436 (1985)

[Ka] Kaufmann, A.G.: Stetigkeit von gruppenwertigen stochastischen Koflächen, Diplomarbeit, Bochum (1987)

[Kl] Klauder, J. R.: Measures and support in functional integration, in *Progress in Quantum Field Theory*, Ed. H. Ezawa, S. Kamefuchi, Elsevier Science, 31–56 (1986)

[Ku] Kusuoka, S.: Markov fields and local operators, J. Fac. Sci. Univ. Tokyo 1A, **26**, 199–212 (1979)

[Lö] Löffelholz, J. : Faradays law and quantum theory, Karl-Marx Universität, Leipzig, Preprint (1987)

[Ne] Nelson, E. : Probability theory and Euclidean field theory. In *"Constructive Quantum Field Theory"*, Edts. G. Veto, A. Wightman, Springer, New York (1973)

[O] Ossipov, E.P. : Euclidean Markov fields from stochastic differential equations in eight-dimensional spaces, Novosibirsk Preprint (1987)

[Rö] Röckner, M. : Traces of harmonic functions and a new path space for the free quantum field, J. Funct. Anal. 79, 211–249 (1988)

[Su1] Surgailis, D.: On infinitely divisible self-similar random fields, Z. Wahrsch. th. verw. Geb. 58, 453–477 (1981)

[Su2] Surgailis, D.: On covariant stochastic differential equations and Markov property of their solutions, Universitá di Roma, Preprint, Fisica (1979)

RANDOM POINT DEFECTS

Point interaction Hamiltonians for crystals with random defects

by

S. Albeverio[1,2,3,7], R. Figari[2,3,4], F. Gesztesy[6], R. Høegh-Krohn[†], H. Holden[5], W. Kirsch[1,3]

Abstract:

We give a short report on work done in recent years on solvable models for quantum mechanical crystals (crystals with point interactions, thus three dimensional extensions of Kronig Penney's model). We discuss the mathematical definition of the Hamiltonian and its spectral properties in the case of perfect crystals, as well as in the case of crystals with deterministic or randomly distributed point defects. We also discuss the connection of such point interactions Hamiltonians with the study of scattering by a large number of small randomly placed scatterers.

1. Introduction

In this paper we shall report on some recent mathematical work on models describing crystals with defects which are randomly distributed. In the formulation of these quantum mechanical models so called point interactions arise; these are interactions localized at points of the perturbed crystal, and are felt by the particle (electron) moving in these crystals (in the usual one-electron approximation of the motions of electrons in a crystal). The crystals we discuss here are mainly three dimensional (but similar results are obtained for two and one dimensional crystals). Despite extensive and very interesting work developed in recent years on point interactions in three dimensions, described e.g. in monographs [AGHKH], [DO], it seems that quite a few physicists, mathematical physicists and mathematicians still believe that point interactions only are possible in one dimension—an immediate association being with the Kronig–Penney model (since it has entered standard text books in solid state physics). This entire workshop has been a proof of how active is the research concerning three dimensional models with point interactions. We hope that the present contribution might also help eliminating eventually the above mentioned prejudice. In fact all is done in one dimension with point interactions can be done also in three dimensions, provided of course the point interactions are correctly defined. We shall report here mainly on work of the authors and their coworkers, in particular, as far as random perturbations are concerned, F. Martinelli. We refer to other contributions in this

---

1. Fakultät für Mathematik, Ruhr-Universität Bochum, D-4630 Bochum 1 (FRG)
2. BiBoS Research Centre
3. SFB 237 Bochum–Essen–Düsseldorf
4. Dipartimento di Scienze Fisiche, Università di Napoli (Italy)
5. Matematisk Institutt, Universitetet i Trondheim, N-7034 Trondheim-NTH (Norway)
6. Department of Mathematics, University of Missouri, Columbia, Mo 65211 (USA)
7. CERFIM Research Centre, Locarno (CH)

† Matematisk Institutt, Universitetet i Oslo (Norway)
Deceased 24 January 1988

volume for complementary topics, see also e.g. [Pa] (and references therein). Our basic reference for this paper is the monography [AGHKH], to which we also refer for more complete references. For the reader at his first contact with point interactions let us start by answering briefly the question:

1.1 What are point interactions ?

A point interaction at the origin 0 in the d-dimensional Euclidean space $\mathbb{R}^d$ should be a perturbation ("potential") localized at 0 of the free Hamiltonian, thus the Hamiltonian (Schrödinger operator) describing this interaction has the form (in suitable units)

$$H = \text{“} -\Delta + \lambda\delta(x) \text{“} \tag{1.1}$$

as an operator in $L^2(\mathbb{R}^d, dx)$ (square integrable functions over $\mathbb{R}^d$). Does H exist as a well defined self-adjoint operator, is it non trivial in the sense of being different from the free part $-\Delta$? More generally point interactions at a subset Y of R^d should be

$$H = \text{“} -\Delta + \sum_{y \in Y} \lambda_y \delta(x-y) \text{“} \tag{1.2}$$

in $L^2(\mathbb{R}^d, dx)$.

Models of this type, with different choices of Y, occur in nuclear physics, solid state physics, electromagnetism, see [AGHKH] and references therein. It is well known that there is a "no go theorem" for Y discrete (without accumulation points) if $d \geq 4$, $-\Delta$ being already essentially self-adjoint on $C_0^\infty(\mathbb{R}^d - \{0\})$ (C^∞ functions of compact support in the complement of the origin) if $d \geq 4$. [1]).

It is also well known that, as an application of Krein's theory, as first discussed by Berezin and Faddeev, when $d \leq 3$ for Y consisting of only one point there exists a 1 parameter family indexed by $\alpha \in \mathbb{R}$, of different realizations $-\Delta_\alpha \neq -\Delta$ of H. The parameter α determines for $d = 2, 3$ a renormalized coupling constant. For $d = 1$, α is simply λ. The way $-\Delta_\alpha$ arises is perhaps best seen by an heuristic argument (justifiable by nonstandard analysis [AFHKL]):

$$\left(-\Delta + \lambda V - k^2\right)^{-1} = G_k - G_k \left[\frac{1}{\lambda} + VG_k\right]^{-1} VG_k\,, \tag{1.3}$$

with $G_k \equiv \left(-\Delta - k^2\right)^{-1}$, $\quad k^2 \neq 0$, as computed rigorously for V say bounded and continuous. Setting then formally $V(x) = \delta(x)$ in this formula one sees that, for $d = 2, 3$, one has to choose $\frac{1}{\lambda} = -G_0(0) - \alpha$ to compensate the singularity of $|G_k(0)|$ ($= +\infty$ for $d = 2, 3$). By this choice of λ we get $\left[\frac{1}{\lambda} + VG_k\right] = \frac{ik}{4\pi} - \alpha$. As suggested by this (1.1) can then be defined using the final result of this formal computation, namely as the selfadjoint operator $-\Delta_\alpha$ in $L^2(\mathbb{R}^d, dx)$, $d \leq 3$ with resolvent kernel given by

$$\left(-\Delta_\alpha - k^2\right)^{-1}(x, y) = G_k(x-y) - G_k(x)\left(\frac{ik}{4\pi} - \alpha\right)^{-1} G_k(y)\,. \tag{1.4}$$

---

[1]) It is a different story if instead of $L^2(\mathbb{R}^d, dx)$ one considers some other spaces, as in some uses of point interactions in electromagnetic theory, see references to work by Grossman and Wu in [AGHKH]. For recent results on point interactions situated on non discrete subsets Y see in addition to [AGHKH] and contributions to these proceedings, [Bra], [ABrR], [AMaZ1-3], [AFHKL], [AFHKKL], [Ko], [H], [Pan], [Te] and references therein.

Remark: The mentioned justification of the above formal computation by nonstandard analysis yields that, for $d = 3$, and ε infinitesimal, $-\Delta + \lambda_\alpha(\varepsilon)\delta_\varepsilon(x)$, with $\delta_\varepsilon(x)$ a nonstandard realization of the δ–function (in standard terms, $\delta_{\tilde\varepsilon}(x)$ is a delta sequence for $\tilde\varepsilon \downarrow 0$), is near standard and defines $-\Delta_\alpha \neq -\Delta$ if $\lambda_\alpha(\varepsilon) = \left[-\frac{\pi^2}{4\varepsilon^2} + \frac{8\pi}{\varepsilon}\alpha\right]\frac{4}{3}\pi\varepsilon^3$, $\alpha \in \mathbb{R}$. A similar result holds also for $d = 2$. See [AFHKL] and references therein. (1.4) or the observation in the remark give a realization of $-\Delta_\alpha$ as a $\delta-$ interaction or point interaction of strength α at the origin.

Remarks 1) It is useful to remark that setting $\alpha = +\infty$ formally in (1.4) the r.h.s. reduces to G_k, so that it is natural to define $-\Delta_{+\infty} \equiv -\Delta$.
2) Besides the mentioned two ways to define $-\Delta_\alpha$, the one by the resolvent and the nonstandard analytic one, there exist other ways to define $-\Delta_\alpha$ e.g.
a)"by boundary conditions": formulated for $d = 3$, $-\Delta_\alpha$ can be characterized as the extension of $-\Delta$ on $C_0^\infty(\mathbb{R}^3 - \{0\})$ functions with domain such that if $f = D(-\Delta_\alpha)$ and $f(x) = \tilde f(r)$, $r \equiv |x|$, then $\frac{\partial}{\partial r}\left(r\tilde f(r)\right) = 4\pi\alpha\left(r\,\tilde f(r)\right)$ at $r = 0$.

b) "by resolvent limits": define for $\varepsilon > 0$, $H_\varepsilon = -\Delta + \lambda(\varepsilon)\varepsilon^{-2}V(x/\varepsilon)$, with V in Rollnik's class, $(1 + |\cdot|)V \in L^1(\mathbb{R}^3)$, and $\lambda \in C^1(\mathbb{R})$, $\lambda(0) = 1$, $\lambda'(0) \neq 0$ such that -1 is an eigenvalue of $v\,\mathrm{sign}V\,G_0 v$, $v \equiv |V|^{\frac{1}{2}}$ with eigenfunctions φ_j, $j = 1, \ldots, N$ in $L^2(\mathbb{R}^3, dx)$ such that the "resonance functions" $G_0 v\varphi_j$ are not in $L^2(\mathbb{R}^3, dx)$ for some j. Then H_ε converges in norm resolvent sense as $\varepsilon \downarrow 0$ to $-\Delta_\alpha$, with $\alpha = -\frac{\lambda'(0)}{\sum_{j=1}^N |(v,\varphi_j)|^2}$ This is an approximation of $-\Delta_\alpha$ by local potentials, there are also approximations by non local potentials, see again [AGHKH].

c) Another useful construction of $-\Delta_\alpha$, which provides probabilistic tools for the study of point interactions, has been first discussed in [AHKS]. Let, for $d = 3$, $\varphi_\alpha(x) \equiv |\alpha|^{\frac{1}{2}}\frac{e^{4\pi\alpha|x|}}{|x|}$, $\alpha \in \mathbb{R}$. Let $\hat H_\alpha$ be the self-adjoint positive operator in $L^2(\mathbb{R}^3, |\varphi_\alpha|^2\,dx)$, uniquely associated with the Dirichlet form $E(f,f) = \frac{1}{2}\int |\nabla f|^2 |\varphi_\alpha|^2 dx$ in $L^2(\mathbb{R}^3, |\varphi_\alpha|^2 dx)$
(in the sense that $\left(\hat H_\alpha^{\frac{1}{2}} f, \hat H_\alpha^{\frac{1}{2}} f\right) = E(f,f)$, $\forall f \in D(\hat H_\alpha^{\frac{1}{2}}) = D(E)$). Then we can define $-\Delta_\alpha$ by $-\Delta_\alpha = \varphi_\alpha \hat H_\alpha \varphi_\alpha^{-1} - (4\pi\alpha)^2$. $\hat H_\alpha$ on $C_0^\infty(\mathbb{R}^3 - \{0\})$ is given by $-\Delta - \beta_\alpha \cdot \nabla$, with $\beta_\alpha \equiv \nabla \ln \varphi_\alpha$. $\hat H_\alpha$ generates a diffusion Markov symmetric semigroup in $L^2(\mathbb{R}^3, |\varphi_\alpha|^2 dx)$, with invariant measure $|\varphi_\alpha|^2 dx$.

Having solved the problem of the construction of self-adjoint realizations of the one source point interaction, by one of the above methods, it is not difficult to extend the solution to other cases, of the type (1.2.) with Y consisting e.g. of N points in $\mathbb{R}^3$ or a discrete subset of $\mathbb{R}^3$ (see below). One can also discuss the case where Y is some other suitable geometrical measure zero subsets of $\mathbb{R}^3$ like e.g. S^2 (see e.g. [AGS]) or the path of Brownian motion on $\mathbb{R}^d, d \leq 5$, run in $[0,t]$ (see [AFHKL], [AFHKKL]). In this article we shall discuss some situations where the particle and the centers are in $\mathbb{R}^3$, for other cases in particular finite and infinitely many centers in $\mathbb{R}^d, d = 1, 2$, see [AGHKH] and also e.g. [DO], as well as contributions to this conference, in particular by P. Exner and P. Šeba.

1.2 *N*-centers point interactions

Before going over to the case of infinitely many centers, let us consider the N-centers case, given heuristically by (1.2.) with Y consisting of N points ("sources") in $\mathbb{R}^3$. We give strengths $\{\alpha_y, y \in Y\}$ corresponding to the sources $y \in Y$. The point interaction Hamiltonian for the sources y, with strengths α_y, denoted by $-\Delta_{\alpha,Y}$, is given in terms of its resolvent by

$$\left(-\Delta_{\alpha,Y} - k^2\right)^{-1} = G_k - \sum_{y,y' \in Y}\left(\overline{G_k(\cdot - y)}, \cdot\right)\left[\Gamma_{\alpha,Y}(k)\right]^{-1}_{yy'} G_k(y' - \cdot), \tag{1.5}$$

with

$$[\Gamma_{\alpha,Y}(k)]_{y,y'} \equiv \left[\left(\frac{ik}{4\pi} - \alpha_y\right)\delta(y-y') + \tilde{G}_k(y-y')\right]_{y,y'}, y, y' \in Y \tag{1.6}$$

where $\tilde{G}_k(z) = G_k(z)$ for $z \neq 0$ and $\tilde{G}_k(z) = 0$ for $z = 0$.

Remark. Actually there exist N^2 self-adjoint extensions of $-\Delta$ restricted to C^∞ funtions of compact support outside the sources $y \in Y$, only N of which are covered by above definition. However only the ones given above do correspond to separated boundary conditions at the sources, the others are "non local" involving non separated boundary conditions, see [DaGr], [Bra]. Having the above resolvent it is possible to discuss in details spectral properties and scattering for N centers, see [AGHKH].

1.3. The case of infinitely many centers

We shall consider the case of Hamiltonians given formally by (1.2) with Y an infinite subset of $\mathbb{R}^3$, discrete in the sense that

$$Y = \{y_i, i \in \mathbb{N}\} \text{ with } \inf_{j\neq j'} |y_j - y_{j'}| > 0.$$

We give sources α_j, $j \in \mathbb{N}$ and denote for any $\tilde{Y} \subset Y$, $\tilde{Y}$ finite, by $\tilde{\alpha}$ the restriction of α to $\tilde{Y}$. One then defines the Hamiltonians for point interactions with sources Y and strengths α by the limit in the sense of strong resolvent convergence of $\left(-\Delta_{\tilde{\alpha},\tilde{Y}} - k^2\right)^{-1}$ as $\tilde{Y} \uparrow Y$.

That this limit exists is easily seen by using monotonicity arguments, see [AGHKH]. Since by (1.2) we have good control on the approximants it is possible to get information on the limit, especially in the case where (α, Y) have suitable symmetry properties, see [AGHKH]. Such a case is the one of crystals, which we shall handle in the next chapter.

2. Periodic point interactions and crystals.

We shall consider in the one electron model of a solid the case of a multiatomic crystal or a perfect alloy, with point sources located at the points of a subset Y of $\mathbb{R}^3$ of the form

$$Y = \Lambda + Y_0, \text{ with } \Lambda \equiv \left\{\sum_{i=1}^{3} n_i a_i, (n_1, n_2, n_3) \in \mathbb{Z}^3\right\}$$

a Bravais lattice and Y_0 a finite number of points of $\mathbb{R}^3$. Let $\hat{\Gamma} \equiv \mathbb{R}^3/\Lambda$ be the basic periodic or primitive cell or Wigner-Seitz cell, i.e.

$$\hat{\Gamma} \equiv \left\{\sum_{i=1}^{3} s_i a_i, s_i \in \left[-\frac{1}{2}, \frac{1}{2}\right)\right\}.$$

Let b_j be dual basis vectors in the sense that

$$a_i b_j = 2\pi\delta_{ij}, \quad i, j = 1, 2, 3.$$

Let

$$\Gamma \equiv \left\{\sum_{i=1}^{3} n_i b_i, (n_1, n_2, n_3) \in \mathbb{Z}^3\right\}.$$

Γ is called the dual lattice.

$$\hat{\Lambda} \equiv \mathbb{R}^3/\Gamma \equiv \left\{ \sum_{i=1}^{3} s_i b_i, s_i \in \left[-\frac{1}{2}, \frac{1}{2}\right) \right\}$$

is the so called Wigner-Seitz cell of the dual lattice Γ or Brillouin zone. $\hat{\Gamma}$ is the dual group of Λ and is the basic periodic cell or primitive cell of the dual lattice. One has the direct decomposition $L^2\left(\hat{\Lambda}, l^2(\Gamma)\right) = \int_{\hat{\Lambda}}^{\oplus} l^2(\Gamma) d\theta$. We can look upon $\theta + \gamma, \gamma \in \Gamma$ as coordinates corresponding to $\hat{p} \in \mathbb{R}^3$. Our periodic Hamiltonian $\hat{H}$, describing crystals (in Fourier space), is unitarily equivalent to a direct integral:

$$\hat{H} \cong \int_{\hat{\Lambda}}^{\oplus} \hat{H}(\theta) d\theta,$$

for some $\hat{H}(\theta)$ acting in $l^2(\Gamma)$.
The study of the spectrum of $\hat{H}$ is then reduced to the study of the spectrum of $\hat{H}(\theta)$.
If $\hat{H}$ is the momentum space realization of a Hamiltonian, then $\hat{H}(\theta)$ is called the reduced Hamiltonian. E.g. if H is $-\Delta$, then $-\hat{\Delta}(\theta)$ is the operator of multiplication by $|\gamma+\theta|^2$ in $l^2(\Gamma)$ and the spectrum of $-\hat{\Delta}(\theta)$ is the discrete set $|\Gamma+\theta|^2$, so that $\sigma(-\Delta) = \bigcup_{\theta \in \hat{\Lambda}} \sigma(-\hat{\Delta}(\theta)) d\theta$ consists of bands, the spectrum in each band being purely absolutely continuous. We shall now discuss the Hamiltonians corresponding to point interactions located at the points $Y_0 + \Lambda$. Formally it is given by $-\Delta + V(x)$, with

$$V(x) = - \sum_{\substack{y_j \in Y_0 \\ j=1,\dots,N}} \sum_{\lambda \in \Lambda} \mu_j \delta(x - y_j - \lambda), \tag{2.1}$$

with (unrenormalized) strengths $\mu_j \in \mathbb{R}$, $j = 1, \dots, N$ (with N number of points in Y_0).
The following Theorem, proved in [AGHKH], shows how to construct the point interaction Hamiltonian corresponding to the interaction given by the above formal expression (2.1):

<u>Theorem 2.1</u> Let

$$\left(\hat{H}^K(\theta) g\right)(\gamma) \equiv |\gamma+\theta|^2 g(\gamma) - \frac{1}{|\hat{\Gamma}|} \sum_{j=1}^{N} \mu_j^k \left(\phi_{y_j}^K(\theta), g\right) \phi_{y_j}^K(\theta),$$

with $K > 0$ a cut-off, (,) the scalar product on $l^2(\Gamma)$, and

$$\phi_{y_j}^K(\theta, \gamma) \equiv \chi_{_K}(\gamma+\theta) e^{-i(\gamma+\theta) y_j},$$

χ_K being the characteristic functions of the closed ball of radius K centered at the origin. We have that

$$D\left(\hat{H}^K(\theta)\right) = D\left(-\hat{\Delta}(\theta)\right) = \left\{ g \in l^2(\Gamma) |, \sum_{\gamma \in \Gamma} |\gamma+\theta|^4 |g(\gamma)|^2 < \infty \right\}.$$

Choose $\mu_j^K \equiv \left(\alpha_j + \frac{K}{2\pi^2}\right)^{-1}$, with $\alpha_j \in \mathbb{R}$. Then $\hat{H}^K(\theta)$ converges in norm resolvent sense as $K \longrightarrow \infty$ to a self-adjoint operator $-\hat{\Delta}_{\alpha,\Lambda,Y_0}(\theta)$, the reduced Hamiltonian for point interactions on $\Lambda + Y_0$. This operator is given by its resolvent through the formula

$$\left(-\hat{\Delta}_{\alpha,\Lambda,Y_0}(\theta) - k^2\right)^{-1} = G_k(\theta) + |\hat{\Gamma}|^{-1} \sum_{j,j'=1}^{N} \left[-\Gamma_{\alpha,\Lambda,Y_0}(h,\theta)\right]_{jj'}^{-1} \left(F_{-\bar{k},y_{j'}}(\theta,\cdot),\cdot\right) F_{k,y_j}(\theta,\cdot),$$

$$[\Gamma_{\alpha,\Lambda,Y_0}(k,\theta)]_{j,j'} \equiv \alpha_j \delta_{jj'} - g_k(y_j - y_{j'},\theta),$$

with

$$g_k(x,\theta) \equiv \begin{cases} \frac{1}{|\hat{\Gamma}|} \lim_{K\to\infty} \sum_{\gamma\in\Gamma, |\gamma+\theta|\le K} \frac{e^{i(\gamma+\theta)x}}{|\gamma+\theta|^2 - k^2} & \text{for } x \in \mathbb{R}^3 - \Lambda \\ (2\pi)^{-3} e^{-i\theta x} \lim_{K\to\infty} \left[-4\pi K + \sum_{\gamma\in\Gamma|\gamma+\theta|\le K} \left(|\gamma+\theta|^2 - k^2\right)^{-1} |\hat{\Lambda}| \right] & \text{for } x \in \Lambda. \end{cases}$$

$$F_{k,y}(\theta,\gamma) \equiv \frac{e^{-i(\gamma+\theta)y}}{|\gamma+\theta|^2 - k^2}, G_k(\theta) \equiv \left(|\gamma+\theta|^2 - k^2\right)^{-1}.$$

The periodic point interaction Hamiltonian with sources on $\Lambda + Y_0$ and strengths $\alpha_{y_j} = \alpha_{y_j+\Lambda}$ (independent of $\lambda \in \Lambda$) is given by $-\Delta_{\alpha,\Lambda+Y_0} \cong \int_{\hat{\Lambda}}^{\oplus} \left(-\Delta_{\alpha,\Lambda,Y_0}(\theta)\right) d\theta$, hence determined by the above resolvent. ■

In the simplest case $Y_0 = \{0\}$, writing $-\hat{\Delta}_{\alpha,\Lambda}$ for $-\hat{\Delta}_{\alpha,\Lambda+\{0\}}$, we have the following spectral results:

Theorem 2.2 $\sigma\left(-\hat{\Delta}_{\alpha,\Lambda}(\theta)\right)$ is purely discrete consisting of isolated eigenvalues of finite multiplicity. We have $\mathbb{R} - |\Gamma+\theta|^2 = \bigcup_{n=0}^{\infty} I_n(\theta)$, with $I_0(\theta) = (-\infty, \theta^2)$, $I_n(\theta)$ being bounded disjoint open intervals for $n \ge 1$, each containing exactly one eigenvalue $E_n^{\alpha,\Lambda}(\theta)$ for $-\hat{\Delta}_{\alpha,\Lambda}(\theta)$. These eigenvalues are increasing in α. In addition a point $E^{\Lambda}(\theta) \in |\Gamma+\theta|^2$ is an eigenvalue of $-\hat{\Delta}_{\alpha,\Lambda}(\theta)$ of multiplicity $m \ge 1$ if and only if there exist $\gamma, ..., \gamma_m \in \Gamma$ such that $E^{\Lambda}(\theta) = |\gamma_0+\theta|^2 = ... = |\gamma_m+\theta|^2$.

For the proof see [AGHKH]. Having this result one then gets information on the spectrum of the point interaction Hamiltonian for a crystal Λ:

Theorem 2.3 The spectrum $\sigma(-\Delta_{\alpha,\Lambda})$, of a crystal with point interactions on a lattice Λ, with strengths α equal at each point of the lattice, is purely absolutely continuous and has the form of the union of two intervals

$$\sigma\left(-\Delta_{\alpha,\Lambda}\right) = \left[E_0^{\alpha,\Lambda}(0), E_0^{\alpha,\Lambda}(\theta_0)\right] \cup \left[E_1^{\alpha,\Lambda}, \infty\right),$$

with

$$\theta_0 \equiv -\frac{1}{2}(b_1 + b_2 + b_3),$$

$$E_1^{\alpha,\Lambda} \equiv \min\left\{E_{b_-}^{\alpha,\Lambda}(0), \frac{1}{4}|b_-|^2\right\} = \min_{\theta\in\hat{\Lambda}} \left[E_{b_-}^{\alpha,\Lambda}(\theta)\right],$$

with

$$|b_-| \le b_j, j = 1,2,3, \quad b_- \in \{b_1, b_2, b_3\}.$$

We have $E_1^{\alpha,\Lambda} > 0$ for all $\alpha \in \mathbb{R}$. Moreover $E_0^{\alpha,\Lambda}(\theta_0) < 0$ if $\alpha < \alpha_{0,\Lambda} \equiv g_0(0,\theta_0)$ (in this case we have thus effectively a gap!). The spectrum of $\sigma\left(-\Delta_{\alpha,\Lambda}\right)$ is monotone increasing in α. One has

$$E_0^{\alpha,\Lambda}(0) \longrightarrow \begin{cases} 0 & \text{as } \alpha \longrightarrow +\infty \\ -\infty & \text{as } \alpha \longrightarrow -\infty \end{cases}$$

and

$$E_0^{\alpha,\Lambda}(\theta_0) \longrightarrow \begin{cases} |\theta_0|^2 & \text{as } \alpha \longrightarrow +\infty \\ -\infty & \text{as } \alpha \longrightarrow -\infty \end{cases}$$

There exists an $\alpha_{1,\Lambda} \in \mathbb{R}$ such that

$$\sigma\left(-\Delta_{\alpha,\Lambda}\right) = \left[E_0^{\alpha,\Lambda}(0), \infty\right) \qquad \forall \alpha \ge \alpha_{1,\Lambda}$$

(i.e. for large enough strengths the gap closes).

Remark. a) There exist some extensions of this result to the case where the basic cell consists of more than one point i.e. $|Y_0| > 1$. E.g. in such a case $\sigma(-\Delta_{\alpha,Y_0+\Lambda}) \cap (-\infty, 0)$ consists of at most $|Y_0|$ disjoint closed bands, see [AGHKH] and references therein.
b) One can approximate $-\Delta_{\alpha,Y_0+\Lambda}$ in the norm resolvent sense by scaled short range Hamiltonians, which can be exploited for obtaining information on crystals with interactions not of the point interaction type.

3. Random point interactions, defect crystals.

We consider the models of Sect. 2 with the sources on $Y_0 + \Lambda$ replaced by random sources located at the random subset $Y(\omega)$ of $\mathbb{R}^3$, with ω a point in a probability space Ω, $Y(\omega)$ being for each ω a countable subset $\{y_j(\omega), j \in \mathbb{N}\}$ of $\mathbb{R}^3$ such that $\inf_{j,j' \in \mathbb{N}} |y_j(\omega) - y_{j'}(\omega)| > 0$.
Let $\alpha(\omega) = \{\alpha_{y_j}(\omega), j \in \mathbb{N}\}$ be a Y-indexed family of random strengths (real valued variables). For each ω, by the methods indicated in Sect. 1.3, we can define a point interaction Hamiltonian

$$H_\omega = -\Delta_{\alpha(\omega),Y(\omega)}, \qquad \omega \in \Omega.$$

A particularly interesting case is the following. Let $X_\lambda, \lambda \in \Lambda$ be i.i.d. $\{0,1\}$-valued random variables associated with the points of a Bravais lattice Λ_0. Set $p = P(X_\lambda = 1)$. Choose $Y(\omega)$ to be the set of occupied sites in Λ i.e. $Y(\omega) = \{\lambda \in \Lambda, X_\lambda(\omega) = 1\}$. Assume $\{\alpha_\lambda, \lambda \in Y\}$ are i.i.d. random variables with supp P_{α_0} compact. Then H_ω has the interpretation of a point interaction Hamiltonian describing a crystal with randomly distributed defects. If $\Lambda(\omega) = \Lambda$ then it is natural to talk of a random alloy, with types of alloys described by the state space of α. Using the fact that both α and Y are i.i.d. we have the following theorem:

Theorem 3.1 $(H_\omega, \omega \in \Omega)$ is an ergodic family of self adjoint operators in $L^2(\mathbb{R}^3)$ (relative to the natural shift operator in path space). The spectrum $\sigma(H_\omega)$ and its different parts like $\sigma_{ess}(H_\omega)$, and the closure $\overline{\sigma_p(H_\omega)}$ of the point spectrum of H_ω, are non random subsets of $\mathbb{R}$, almost surely. Moreover, the discrete spectrum $\sigma_d(H_\omega)$ is void, almost surely. Finally, for any $\tau \in \mathbb{R}$ there exists a subset Ω_τ of Ω of probability 1 such that τ is not an eigenvalue of finite multiplicity of H_ω for $\omega \in \Omega_\tau$.

Remark. This result belongs to a type of results established in various contexts by Pastur, Kirsch-Martinelli, Englisch-Kürsten, see e.g. [AGHKH], [Ki], [KiMa] and references therein.

It is useful to refer to the family $\Phi(\omega) \equiv \{\alpha_\lambda(\omega), X_\lambda(\omega), \lambda \in \Lambda\}$ as to a "stochastic potential". Let $H(\Phi(\omega)) \equiv H_\omega \equiv -\Delta_{\alpha(\omega),Y(\omega)}$ be the corresponding Hamiltonian. We call any $\phi \equiv \{(\xi_\lambda, \eta_\lambda) \in \text{supp} P_{\alpha_0} \times \{0,1\}, \lambda \in \Lambda\}$ an admissible potential. The set of all admissible potentials is denoted by $\mathcal{A}$. Let us set $\Lambda(\phi) \equiv \{\lambda \in \Lambda, \eta_\lambda = 1\}$ and $H(\phi) \equiv -\Delta_{\alpha(\phi),\Lambda(\phi)}$. We call $\phi \in \mathcal{A}$ periodic with periods $L_i, i = 1,2,3$ if there exist linearly independent $L_i \in \Lambda - \{0\}$ such that $\xi_{\lambda+L_i} = \xi_\lambda$ and $\eta_{\lambda+L_i} = \eta_\lambda$ for all $\lambda \in \Lambda$ and all $i = 1,2,3$. This means that the charges as well as the occupied sites are L_i-invariant. We call $\mathcal{P}$ the family of all periodic admissible potentials. We then have

Theorem 3.2 The spectrum $\sigma(H(\phi))$ of the Hamiltonian $H(\phi)$ for any admissible potential ϕ is contained in Σ with $\Sigma = \sigma(H(\Phi(\omega)))$ for almost every $\omega \in \Omega$. Moreover $\Sigma = \bigcup_{\phi \in \mathcal{A}} \sigma(H(\phi)) = \overline{\bigcup_{\phi \in \mathcal{P}} \sigma(H(\phi))}$.

Proof: This also follows from the references mentioned in the previous theorem, see e.g. [AGHKH].

Remark: One can show in addition that:
a) Given $p \leq 1$ the set Σ depends only on suppP_{α_0}

b) Σ has a band structure, see [AGHKH].

It is possible to give a rather detailed study of the variation of the negative part of the spectrum of $H(\phi(\omega))$ as one removes some of the point interactions. In fact we have

Theorem 3.3

$$\text{Let } \mu \equiv \inf[\text{supp}P_{\alpha_0}],$$
$$\nu \equiv \sup[\text{supp}P_{\alpha_0}].$$

Assume that either $E_0^{\nu,\Lambda}(0) \leq E_0^{\mu,\Lambda}(\theta_0)$ or supp$P_{\alpha_0} = [\mu,\nu]$. Then the assumption $p = 0$ implies

$$\begin{aligned}\Sigma &= \left[E_0^{\mu,\Lambda}(0), E_0^{\nu,\Lambda}(\theta_0)\right] \cup \left[E_1^{\mu,\Lambda}, \infty\right] \\ &= \sigma(-\Delta_{\mu,\Lambda}) \cup \sigma(-\Delta_{\nu,\Lambda}).\end{aligned} \tag{3.1}$$

If $\nu < \alpha_{0,\Lambda}$ we have

$$\Sigma \cap (-\infty, 0) = \left[E_0^{\mu,\Lambda}(0), E_0^{\nu,\Lambda}(\theta_0)\right]$$

and if $\mu \geq \alpha_{1,\Lambda}$ we have $\Sigma = \sigma(-\Delta_{\mu,\Lambda})$.

Remark. a) If $0 < p < 1$ the same results hold, but (3.1) should be replaced by

$$\Sigma = \left[E_0^{\mu,\Lambda}(0), E_0^{\nu,\Lambda}(\theta_0)\right] \cup [0,\infty) = \sigma(-\Delta_{\mu,\Lambda}) \cup \sigma(-\Delta_{\nu,\Lambda}) \cup [0,\infty).$$

b)For $\mu > \alpha_{1,\Lambda}$ we have the result that $\sigma(H_\omega) = \sigma(-\Delta_{\mu,\Lambda})$ almost surely, i.e. the spectrum does not change if starting from the random Hamiltonian we create or switch off points with arbitrary strength.
c) For results on random Hamiltonians formally given by

$$H(\omega) = -\Delta + \lambda \int_0^t \delta(x - b(s,\omega))ds$$

with $b : [0,t] \longrightarrow \mathbb{R}^d$ a Brownian motion in $\mathbb{R}^d, d \leq 5$ see [AFHKL] [AFHKKL] and references therein. Such Hamiltonians arise in the description of the scattering of a quantum mechanical particle by a polymer. Moreover, via the Feynman-Kac formula, they also enter the construction of polymer measures of the heuristic form

$$\exp\left[-\lambda \int_0^t \int_0^t \delta(b(s_1) - \tilde{b}(s_2))ds_1 ds_2\right] dP(b)d\tilde{P}(\tilde{b}),$$

with (b,P), $(\tilde{b},\tilde{P})$ two independent Brownian motions in $\mathbb{R}^d, d \leq 4$. The densities of such measures also occur in Symanzik's representations of self-interacting scalar quantum fields described by interaction densities $v(s)$ which are functions of s^2, like e.g. in the $(\varphi^4)_d$-model, see [AFHKL].

4. Point interactions and scattering by a large number of small randomly placed scatterers

In the definition of the N center point interaction (1.5.) all the physical constants characterizing the strength of the interaction, the energy of the quantum particle and the mutual distances of the scatterers are contained in the matrix $\Gamma_{\alpha,Y}(k)$.

In particular they appear in Γ as the inverses of three lengths: the $\alpha_y, y \in Y = \{Y_1, ..., Y_N\}$ represent the inverses of an "effective linear size " of the scatterers, (see [AGHKH]), k is proportional to the inverse of the wave length associated with the quantum particle and the $G_k(y - y')$ are proportional to the inverses of the distances between the scatterers. In this section we shall study some limits $N \longrightarrow \infty$, different from those studied in Sect. 2,3, but also of physical relevance cfr e.g. [Lo]. We report here on work in [FHT] - [FOT].

In the following the orders of magnitude of the above lengths will be expressed as powers of the number N of the scatterers and the volume of the system will be considered fixed and finite (of course only the dimensionless ratios of the lengths are to be considered as meaningful parameters). We shall consider separately three situations, which we shall denote by cases 1), 2), 3).

Case 1:

$$\inf_{y,y' \in Y} |y - y'| = 0(N^{-\frac{1}{3}}), k^{-1} = 0(1), \alpha_y^{-1} = 0(N^{-1})$$

. In this case there is a large number N of scatterers in a finite volume, each one of strength being of order N^{-1}. The wavelength of the quantum particle is large with respect to the scatterers effective length and to the interparticle distance.

Physically, in this case one expects a finite effective potential depending only on the local strength per unit volume of the scatterers.

Case 2:

$$\inf_{y,y' \in Y} |y - y'| = 0\left(N^{-\frac{1}{3}}\right), k^{-1} = 0\left(N^{-\frac{1}{3}}\right), N^{\frac{1}{2}} \alpha_y^{-1} = 0(N^{-\frac{1}{3}}).$$

In this case the wavelength is of the same order as the interparticle distance. Each scatterer has infinitesimal strength but the "surface per unit volume" is kept constant.

The limit problem is expected to contain information about the local statistics of the interparticle distances. It is the relevant limit, for example, in modelling scattering of neutrons by liquids (scattering due only to the nuclei of the atoms in the liquid). In fact this kind of scattering experiments are often performed to investigate the range of solid-like order in the interparticle distances in fluids.

Case 3: All the lengths are of order $N^{-\frac{1}{3}}$. This case is in fact a genuine infinite volume limit case. There are no negligible terms in the Γ matrix.

This case is the relevant one for modelling amorphous or perfect crystalline solids, as discussed in Sect. 2,3. In what follows we give results for cases 1) and 2), for dimension $d = 3$ (the case $d = 2$ can be worked out as the $d = 3$ case, whereas the case $d = 1$ is much simpler).

Case 1 Let $Y^{(N)} = \left\{y_1^{(N)}, ..., y_N^{(N)}\right\}$ be a sequence of N-tuples of points in $\mathbb{R}^3$ such that

$$\frac{1}{N} \sum_{i=1}^{N} \delta_{y_i^{(N)}} \underset{N \uparrow \infty}{\overset{w}{\longrightarrow}} \rho(x) dx$$

with $\rho(x) \geq 0$, $\rho(\cdot) \in L^1(\mathbb{R}^3) \cap L^2(\mathbb{R}^3)$, $\int_{\mathbb{R}^3} \rho(x) dx = 1$, the convergence being weak convergence of probability measures on $\mathbb{R}^3$.

Some technical assumptions on the distribution of the $y_i^{(N)}$ will be needed; e.g.

$$\inf_{i,j} \left| y_i^{(N)} - y_j^{(N)} \right| \geq cN^{-\alpha} \text{ for some } \alpha \in \left(\frac{1}{3}, 1\right)$$

$$\frac{1}{N^2} \sum_{\substack{i,j \\ i \neq j}} \frac{1}{\left| y_i^{(N)} - y_j^{(N)} \right|^2} \leq c \qquad \forall N$$

If, for example, the $y_i^{(N)}$ are chosen to be N independent, identically distributed (i.i.d.) random points of $\mathbb{R}^3$ with common distribution density ρ, the above stated assumptions are satisfied by any configuration of a set of measure increasing to 1 when N goes to infinity.
Let $\alpha^{(N)} = \{\alpha_y, y \in Y^{(N)}\}$. We are looking for the existence of a limit operator for the sequence $-\Delta_{N\alpha^{(N)},Y^{(N)}}$ and for an explicit characterization of the limit. If all the $\alpha_j^{(N)}$ are bounded away from 0 and $\inf_{j \neq j'} \alpha_j^N \left| y_j^{(N)} - y_{j'}^{(N)} \right| \gg 1$, as a first order in a perturbation expansion, we get, as $N \longrightarrow \infty$:

$$\left[\Gamma_{N\alpha^{(N)},Y^{(N)}}\right]_{j,j'} \sim N\alpha_j^{(N)} \delta_{j,j'}$$

and

$$\left(-\Delta_{N\alpha^{(N)},Y^{(N)}} - k^2\right)^{-1}(x,y) \sim G_k(x,y) +$$
$$+ \frac{1}{N} \sum_{j=1}^{N} G_k\left(x, y_j^{(N)}\right) G_k\left(y_j^{(N)}, y\right) \left(\alpha_j^{(N)}\right)^{-1} \tag{4.1}$$

If the $\alpha_j^{(N)}$ are chosen to be the values $\alpha\left(y_j^{(N)}\right)$ in $y_j^{(N)}$ of a function which is continuous (outside a set of $\rho(x)dx$ measure 0) and $0 < a \leq |\alpha| \leq b < \infty$, the right hand side of (4.1) converges to

$$G_k(x,y) + \int_{\mathbb{R}^3} G_k(x,z) \frac{\rho(z)}{\alpha(z)} G_k(z,y) dz =$$
$$= (-\Delta - k^2)(x,y) + \left[(-\Delta - k^2)^{-1} \frac{\rho}{\alpha} (-\Delta - k^2)^{-1}\right](x,y) \tag{4.2}$$

(4.2) is the first term of a perturbation expansion of $(-\Delta + \frac{\rho}{\alpha} - k^2)$; up to the first order $-\Delta_{N\alpha^{(N)},Y^{(N)}}$ behaves for large N like $(-\Delta - \frac{\rho}{\alpha})$.
In fact the result is true up to any order in the expansion. One has in particular the following

Theorem 4.1.: Under the assumptions made before

$$s - \lim_{N \uparrow \infty} \left(-\Delta_{N\alpha^{(N)},Y^{(N)}} + \lambda\right)^{-1} = \left(-\Delta - \frac{\rho}{\alpha} + \lambda\right)^{-1} \equiv A_\rho^\lambda$$

for $\lambda > 0$ sufficiently large.
For a detailed proof and further comments see [FHT]. If the $y_i^{(N)}$ are i.i.d. random points with common distribution density ρ the above stated theorem can be looked upon as an operational law of large number. The corresponding central limit theorem can also be proved:

Theorem 4.2.: For any $f, g \in L^2(\mathbb{R}^3)$ the random variable

$$N^{\frac{1}{2}} \left(f, \left[\left(-\Delta_{N\alpha^{(N)},Y^{(N)}} + \lambda\right)^{-1} - A_\rho^\lambda\right] g\right) \equiv \xi^{(N)}(Y^{(N)})$$

converges in distribution when N goes to $+\infty$ to the gaussian random variable ξ^λ with mean 0 and variance.

$$E(\xi^\lambda) = \left(A_\rho^\lambda f A_\rho^\lambda g, \alpha^{-2} A_\rho^\lambda f A_\rho^\lambda g\right)_{L_\rho^2} - \left(A_\rho^\lambda f, \alpha^{-1} A_\rho^\lambda g\right)_{L_\rho^2}^2$$

(here $L_\rho^2 = L^2(\mathbb{R}^3, \rho dx)$).
For the proof and further comments see [FHT] and [FOT].

Remark: If V is any function in $L^1(\mathbb{R}^3) \cap L^2(\mathbb{R}^3)$ and we take ρ, α as

$$\rho(x) = |V(x)| / \int_{\mathbb{R}^3} |V(x)| dx$$

$$\alpha(x) = (\text{sign } V)(x) / \int_{\mathbb{R}^3} |V(x)| dx$$

so that $\frac{\rho}{\alpha} = V$, then Theorems 4.1, 4.2 express the fact that any one particle Hamiltonian with a potential V of class $L^1 \cap L^2$ can be arbitrarily well approximated by a Hamiltonian with zero range potential on an increasing number of points.

Case 2: In experiments of neutron scattering, neutrons with a wavelength of a few Ångstrøms are used to investigate samples of condensed matter. The average interparticle distances are of the same order of the wavelength, while the range of the interaction of the neutrons with the nuclei of the atoms is of the order
10^{-13} cm.

A first order expansion for the differential cross section $d\sigma/d\Omega$ for a formal potential $2\pi b \sum_{i=1}^{N} \delta(x - y_i^{(N)})$ (in the usual units such that $\hbar = m = 1$) gives

$$\left(\frac{d\sigma}{d\Omega}\right)(x) = b^2 \sum_{i,j}^{N} e^{ix(y_i^{(N)} - y_j^{(N)})}$$

where x is the transferred momentum and $y_i^{(N)}$ are the positions of the scatterers.

Notice that the formal expansion cannot be continued beyond the first order term since infinities due to the singularity of the G_k at coinciding points would appear in each higher order term. For the N-centers point interaction the above mentioned result is exact for N large in the scaling described above as pertaining to case 2 (which in a box of volume proportional to N can be redefined as $|y_i - y_j| = 0(1), k^{-1} = 0(1), N^{\frac{1}{3}} \alpha_j^{-1} = 0(1)$). In fact let $\left(\frac{d\sigma}{d\Omega}\right)^{\frac{1}{2}}_{N_\alpha^{\frac{1}{3}}, Y^{(N)}}$ be the differential cross section corresponding to N-center zero range interactions, all of the same strength $N^{-\frac{1}{3}} \alpha^{-1}$, placed at the points $\{y_1^{(N)}, \ldots, y_N^{(N)}\}$. Under some technical assumptions on the distribution of the $y_i^{(N)}$, it is possible to prove that

$$\lim_{N \nearrow \infty} \left| \left(\frac{d\sigma}{d\Omega}\right)_{N^{\frac{1}{3}} \alpha, Y^{(N)}} - \frac{1}{N\alpha^2} \sum_{i,j} e^{ik\left(y_i^{(N)} - y_j^{(N)}\right)} \right| = 0 \tag{4.3}$$

For this see [DFZ].

Notice that if the $y_i^{(N)}$ are distributed according to an homogeneous point process in $\mathbb{R}^3$ of density ρ and if the static pair correlation function given by

$$\frac{g(r)}{4\pi r^2} dr = Pr\left\{ \left| y_i^{(N)} - y_j^{(N)} \right| \in (r, r + dr) \middle| y_i^{(N)} \right\}$$

is decaying fast enough to ρ for large r, the common limit of the two quantities appearing in (4.3) is $\frac{1}{\alpha^2} + \frac{\rho}{\alpha^2} \hat{g}(x)$ ($\hat{g}$ denotes the Fourier transform of g).

It should be stressed that for the N-centers point interaction the terms of the perturbation expansion, disappearing in the limit $N \longrightarrow \infty$, are explicitly known.

Acknowledgements

This report is based on a lecture given by the first named author when he was visiting Dubna with Raphael Høegh-Krohn in the fall '87. On January 24th, 1988, Raphael suddenly passed away. He had been a standing source of inspiration for all of us and we deeply mourn his departure. We thank J. Brasche, G.F. Dell'Antonio, P. Exner, P. Šeba, W. Karwowski and L. Streit for many interesting and simulating discussions. The kind invitation of the first and fourth author to the Dubna Conference is gratefully acknowledged.

References

[ABrR] S. Albeverio, J. Brasche, M. Röckner, Dirichlet forms and generalized Schrödinger operators, in preparation, to appear in "Lectures on Schrödinger Operators", from the Nordic Summer School in Mathematics 1988, Ed. A. Jensen, H. Holden, Lect. Notes Phys., Springer 1989

FHKKL] S. Albeverio, J.E. Fenstad, R. Høegh-Krohn, W. Karwowski, T. Lindstrøm, Schrödinger operators with potentials supported by null sets, in preparation, to appear in Proc. Symp. in Memory of R. Høegh-Krohn

[AFHKL] S. Albeverio, J.E. Fenstad, R. Høegh-Krohn, T. Lindstrøm, Nonstandard Methods in Stochastic Analysis and Mathematical Physics, Academic Press, Orlando (1986)

[AGHKH] S. Albeverio, F. Gesztesy, R. Høegh-Krohn, H. Holden, Solvable Models in Quantum Mechanics, Springer, New York (1988)

[AGS] J.P. Antoine, F. Gesztesy, J. Shabani, Exactly solvable models for sphere interactions in quantum mechanics, J. Phys. A 20, 3687-3712 (1987)

[AHKS] S. Albeverio, R. Høegh-Krohn, L. Streit, Energy forms, Hamiltonians and distorted Brownian paths, J. Math. Phys. 18, 907-917 (1977)

[AMaZ1] S. Albeverio, Ma Zhiming, Nowhere Radon smooth measures, perturbations of Dirichlet forms and singular quadratic forms, in preparation, to appear Proc. Bad Honnef Conference, Ed. Christopeit et al. (1988)

[AMaZ2] S. Albeverio, Ma Zhiming, Additive functionals, smooth nowhere Radon and Kato class measures associated with Dirichlet forms, in preparation

[AMaZ3] S. Albeverio, Ma Zhiming, On the perturbations of Dirichlet forms, in preparation

[Bra] J. Brasche, Perturbations of self-adjoint operators supported by null sets, Ph.D. Thesis, Bielefeld (1988), and papers in preparation

[CaS] C. Carvalho, L. Streit, in preparation

[DaGr] L. Dabrowski, H. Grosse, On nonlocal point interactions in one, two and three dimensions, J. Math. Phys. 26, 2777-2780 (1985)

[DFZ] D. Dürr, R. Figari, N. Zanghi, in preparation

[DO] Y.N. Demkov, V.N. Ostrovskii, The Use of Zero-Range Potentials in Atomic Physics (in Russian), Nauka, Moscow 1975; transl. Plenum Prss

[FHT] R. Figari, H. Holden, S. Teta, A law of large numbers and a central limit theorem for the Schrödinger operator with zero-range potentials, J. Stat. Phys. 51 (1988) 205-214

[FOT] R. Figari, E. Orlandi, S. Teta, The Laplacian in regions with many small obstacles, - fluctuations around the limit operator, J. Stat. Phys. 41, 465-487 (1985)

[H] Herczyński, J., On Schrödinger operators with distributional potentials, Warsaw Preprint

[Ki] W. Kirsch, contribution to "Lectures on Schrödinger Operators", from the Nordic Summer School in Mathematics 1988, Ed. A. Jensen, H. Holden, Lect. Notes Phys., Springer 1989

[KiMa] W. Kirsch, F. Martinelli, Some results on the spectra of random Schrödinger operators and their applications to random point interactions, pp. 223-244 in "Stochastic Methods in quantum theory and stochastic mechanics", Ed. S. Albeverio, Ph. Comte, M. Sirugue-Collin, Lect. Notes Phys. **173**, Springer Berlin (1982).

[Ko] Koshmanenko, V.D., Singular perturbations defined by forms, BiBoS-Preprint, July 1988.

[Lo] Lovesey, Theory of Neutron Scattering from Condensed Matter. Vol. 1. Nuclear Scattering, Vol. 2. Polarization Effects and Magnetic Scattering, Clarendon Press, Oxford 1984.

[Pa] B.S. Pavlov, The theory of extensions and explicitely soluble models, Rus.. Math. Surv. 42, 127 - 168 (1987)

[Pan] Pantić, D., Stochastic calculus on the distorted Brownian motion, Belgrade Preprint (1986)

[Te] Teta, A., Quadratic forms for singular perturbations of the Laplacian, SISSA Preprint, in preparation

SCATTERING ON A RANDOM POINT POTENTIAL

B.S.Pavlov, A.E.Ryzhkov
Physical Institute, Leningrad State University, Ulyanovskaya St. 1, Petrodvoretz, 198904 Leningrad, USSR

1. Introduction

The fact that the Schrödinger equation with a zero-range potential can lead to explicitly solvable models was discovered in 1934 when Fermi formulated his famous deuteron model. Only thirty years later, however, an exact mathematical meaning was given to the Fermi heuristic pseudopotential in the work of Berezin and Faddeev [1]. They showed that the corresponding Hamiltonian is nothing else then a self-adjoint extension of a suitably defined symmetric operator (see [2] for a recent review).

It has been shown by one of the present authors [3] that the structure of the standard point interaction models can be enriched substantially when the self-adjoint extensions are constructed in a larger Hilbert space. This idea yields various models of zero-range interaction with an additional internal structure.

In our previous paper [4] this method has been used to construct an explicitly solvable model of scattering on a point object whose internal structure depends on a stochastic process. In the present paper, investigation of the model is continued. We carry out the spectral analysis of the averaged Hamiltonian which turns out to be a dissipative operator. We show that its spectrum contains a complex absolutely continuous branch. We show, moreover, that the model is explicitly solvable. All spectral properties of the averaged Hamiltonian can be derived from an algebraic "dispersion" equation. We construct also the averaged scattering matrix and investigate its properties in the limit when the velocity of the stochastic process tends to zero. We compare the obtained expressions to the "deterministic" S-matrices which correspond to the fixed stochastic states of the system.

The paper is organized as follows. In Section 2 we construct the Hamiltonian describing a Schrödinger particle interacting with a point-like object whose internal structure depends on a two-state Markov process. We investigate also the evolution operator corresponding to this stochastic Hamiltonian. Averaging this evolution operator with respect to trajectories of the Markov process with fixed initial and final states, we get a strongly continuous semigroup. This semigroup acts in the quantum-stochastic space $\mathcal{H}$ which is the tensor product of the space $L^2(\mathbb{R}^3) \oplus \mathbb{C}^2$ of quantum states and the space $\mathbb{R}^2$ of states of the stochastic process under consideration. Its generator $\hat{\mathcal{L}}_\varkappa$ turns out to be a dissipative operator which is nearly normal in the sense that the commutator of resolvents corresponding to $\hat{\mathcal{L}}_\varkappa$ and its adjoint operator $\hat{\mathcal{L}}^*_\varkappa$ is a finite-rank operator only. This fact makes it possible to express spectral properties of $\hat{\mathcal{L}}_\varkappa$ in explicit form.

In Section 3 we formulate an eigenfunction-expansion theorem for the operator $\hat{\mathcal{L}}_\varkappa$. The theorem can be proved using the explicit formulae for the resolvent of $\hat{\mathcal{L}}_\varkappa$ which have been obtained in [4]. The concluding Section is devoted to construction and investigation of the averaged scattering operator $\bar{S}$ which is of the form

$$\operatorname*{s-lim}_{t \to +\infty} \exp(i\hat{\Delta} t)\ J \exp(2i\hat{\mathcal{L}}_\varkappa t)\ J^* \exp(i\hat{\Delta} t) = \bar{S} \ , \tag{1}$$

where $\hat{\Delta}$ is the unperturbed Hamiltonian and J is the corresponding identification operator. We derive there the limit corresponding to "freezing" of the stochastic variable; the obtained exppression for limiting scattering amplitude is compared to the "deterministic" scattering amplitudes f^+, f^- corresponding to fixed stochastic states

$$\lim_{\varkappa \to 0} f_\varkappa = 1/4\ (f^+ + f^-).$$

The factor $1/4$ appears because the contribution corresponding to the relaxation (complex) branch of the absclulely continuous spectrum of $\hat{\mathcal{L}}_\varkappa$ vanishes. This is a consequence of the S-matrix definition (1) and of the fact that the contribution of the real spectral branch to the scattering amplitudes f^+ and f^- is always non-zero. An alternative definition of the S-matrix (based on the corresponding wave operators) gives (after freezing the stochastic factor) a classical limit corresponding to the situation without stochasticity.

Our model can describe, for instance, a situation when a neutron beam is bombarding a massive nucleus localized in a magnetic substance with a small dispersion. It was shown in [5] that for high enough temperature the magnetic momentum $\vec{M}(t)$ of each domain of the magnetic substance conserves its modulus

$$|\vec{M}| = \text{const}$$

and suffers a Brownian motion on a sphere of a constant radius. If the substance is magnetized along one axis and the temperature is not too high, this motion can be considered as a Markov process with two states. We suppose that the nucleus has only two internal states and that the corresponding Hamiltonian can be expressed as a sum of two terms:

$$H_0 + \gamma\langle\vec{\sigma}, \vec{M}(t)\rangle \quad ,$$

where γ is the so-called gyromagnetic factor of the nucleus, $\vec{\sigma}$ is the vector whose components are Pauli matrices

$$\sigma_1 = \begin{pmatrix} 0 & 1 \\ 1 & 0 \end{pmatrix}, \quad \sigma_2 = \begin{pmatrix} 0 & i \\ -i & 0 \end{pmatrix}, \quad \sigma_3 = \begin{pmatrix} 1 & 0 \\ 0 & -1 \end{pmatrix};$$

$$H_0 = \operatorname{diag}\{\lambda_0, \lambda_1\} \quad ,$$

where λ_0, λ_1 are the two possible "levels" of nucleus.

The motion of the magnetic momentum $\vec{M}(t)$ causes shifts of the nuclear "levels" λ_0, λ_1, which become time dependent. In this situation one can investigate the problem of the averaged (with respect to the magnetic momentum trajectories) elastic scattering on the nucleus. To solve this problem, we shall neglect from the very beginning the direct interaction of the neutron beam with the magnetic field. It is obvious that this technical assumption should be removed in a more detailed analysis where the Laplacian must be replaced by a Pauli operator which includes the direct interaction of the neutron magnetic momentum with the substance magnetic momentum $\vec{M}(t)$.

Let us remark that the basic features of the proposed model remain valid also in other stochastic scattering problems.

2. Construction of the model

We start with the following quantum Hamiltonian of the (neutron + nucleus)-system

$$H = (-\Delta) \oplus \{H_0 + \gamma \langle \vec{\sigma}, \vec{M}(t) \rangle\},$$

where the Laplacian $(-\Delta)$ acts in the space $L^2(\mathbb{R}^3)$ and the internal nuclear Hamiltonian $H_0 + \gamma\langle\vec{\sigma}, \vec{M}(t)\rangle$ acts in $\mathbb{C}^2$ (recall that $H_0 = \mathrm{diag}\{\lambda_0, \lambda_1\}$). We describe the substance magnetic momentum $M(t)$ by a Markov process with two states $\begin{pmatrix} M \\ 0 \end{pmatrix}$ and $\begin{pmatrix} 0 \\ M \end{pmatrix}$ corresponding to the magnetic momentum $\vec{M}$ directed up or down along the z-axis, respectively. Supposing for simplicity that $\gamma = 1$, we get the following expression of the internal Hamiltonian

$$A(\vec{M}) = H_0 \pm \begin{pmatrix} M & 0 \\ 0 & -M \end{pmatrix}, \quad \vec{M} = \pm M\vec{e}_z .$$

For a fixed time t, the interaction between the external and internal channels is constructed using the same method as in [3]. We restrict the Laplacian $(-\Delta) \longmapsto (-\Delta_0)$ to a linear set of W_2^2-smooth functions which equal zero in a neighbourhood of the point $x = 0$. Constructing then the corresponding adjoint operator $(-\Delta_0^*)$ we find that its domain is given by

$$D(-\Delta_0^*) = \left\{ u(x) = \frac{u^-}{4\pi|x|} + u^+ + u_0;\ u_0(x) \in D(-\Delta_0),\ u^\pm \in \mathbb{C} \right\}. \tag{2}$$

The boundary form of the operator $(-\Delta_0^*)$ is given by

$$\langle(-\Delta_0^*)u, v\rangle - \langle u, (-\Delta_0^*)v\rangle = u^+\overline{v^-} - u^-\overline{v^+}, \tag{3}$$

where v^+, v^- correspond to the decomposition of the vector $v \in D(-\Delta_0^*)$,

$$v(x) = \frac{v^-}{4\pi|x|} + v^+ + v_0(x) \quad \text{with} \quad v_0(x) \in D(-\Delta_0).$$

We restrict also the internal operator $A(\vec{M}) \longmapsto A_\varphi(\vec{M})$ in such a way that the generating vector $\varphi = \mathrm{const}\cdot\begin{pmatrix} 1 \\ 1 \end{pmatrix}$ turns out to be its deficiency element: $A_\varphi^*(\vec{M})\varphi = i\varphi$. Introducing a "real basis"

$$w^- = (A - iI)^{-1}A\varphi, \qquad w^+ = (A - iI)^{-1}\varphi, \tag{4}$$

we can write an arbitrary element from the domain of A_φ^* in the form [3]

$$\eta = \tilde{\eta} + \eta^- w^- + \eta^+ w^+, \qquad \tilde{\eta} = (A-iI)^{-1}\psi \in D(A_\varphi). \tag{5}$$

Here ψ is perpendicular to φ, i.e., $\psi = \mathrm{const}\cdot\begin{pmatrix}1\\-1\end{pmatrix}$. The boundary form of the operator A_φ^* reads

$$\langle A_\varphi^*\eta, \zeta\rangle - \langle \eta, A_\varphi^*\zeta\rangle = \eta^+\overline{\zeta^-} - \eta^-\overline{\zeta^+}. \tag{6}$$

The operator $(-\Delta_0) \oplus A_\varphi$ has deficiency indices (2,2). It has a four-parameter family of self-adjoint extensions which is parameterized by Lagrange planes of the global boundary form (3) + (6). In our model, we choose the extension $\mathcal{L}$ specified by the following boundary conditions

$$u^- = \alpha\eta^+, \qquad \eta^- = \alpha u^+, \qquad \mathrm{Im}\,\alpha = 0. \tag{7}$$

The operator $\mathcal{L}$ obtained in such a way describes the Hamiltonian of the (neutron + nucleus)-system. Since $\vec{M} = \vec{M}(t)$, this Hamiltonian is time-dependent. The corresponding evolution operator $U(t)$ restricted to a fixed trajectory of the process $\vec{M} = \vec{M}(t)$ is a solution to the equation

$$\frac{1}{i}\frac{\partial U}{\partial t} = \mathcal{L}(\vec{M}(t))U, \qquad U\big|_{t=0} = I_q = I_e \oplus I_i, \tag{8}$$

where I_e and I_i are the identity operators in the external and internal spaces, respectively.

Together with the stochastic evolution described by the equation (8), we shall consider the "deterministic" evolutions corresponding to the Hamiltonians $\mathcal{L}(+M)$, $\mathcal{L}(-M)$ in which the magnetic momentum is fixed in the up-state or in the down-state. On the time intervals where $\vec{M}(t)$ is constant, the evolution equation (8) is solved by the time-ordered exponentials [4] corresponding to the operators $\mathcal{L}(+M)$ and $\mathcal{L}(-M)$, respectively.

In order to calculate the averaged evolution operator we must introduce a measure on the trajectories of the stochastic process $\vec{M}(t)$. The most simple way how to do it is to use the equation for the transition probabilities $p = (p^+(t), p^-(t))$, the solving matrix $\mathcal{P}$ of which represents a solution of the following equation:

$$\frac{d\mathcal{P}}{dt} = \varkappa\begin{pmatrix}-1 & 1\\ 1 & -1\end{pmatrix}, \qquad \mathcal{P}(0) = \begin{pmatrix}1 & 0\\ 0 & 1\end{pmatrix}. \tag{9}$$

Knowing the probabilities $p^+(0)$, $p^-(0)$ of the up- and down-orientations of $\vec{M}(0)$, we can calculate the corresponding probabilities at an arbitrary instant

$$\begin{pmatrix} p^+(t) \\ p^-(t) \end{pmatrix} = \mathcal{P}(t) \begin{pmatrix} p^+(t) \\ p^-(t) \end{pmatrix} .$$

The equation (9) is called equation of stochastic evolution. It allows to calculate the probability of the beam of trajectories of $M(t)$ which are in the states $\alpha_s = \pm M$ at the instants $t = s\Delta$, $s = 0,1,\ldots,n$:

$$P_{\alpha_n \alpha_{n-1} \ldots \alpha_0} = \prod_{s=1}^{n} \left\{ \exp\left[\varkappa \begin{pmatrix} -1 & 1 \\ 1 & -1 \end{pmatrix} \Delta \right] \right\}_{\alpha_s \alpha_{s-1}} .$$

In this way the measure on the space of trajectories can be introduced. After that, one can calculate the averaged evolution operator on the set of trajectories with fixed initial and final states using the Trotter formula. It was carried out in [4] on a "physical level" of rigour. A complete proof for the case of smooth interactions can be found in [6].

Let us consider the quantum-stochastic space $\mathcal{H}$, introduced above, i.e., tensor product of $L^2(\mathbb{R}^3) \oplus \mathbb{C}^2 = \mathcal{H}_q$ with the space of the stochastic states $\mathbb{R}^2$, $\mathcal{H} = \mathcal{H}_q \oplus \mathcal{H}_q$. Consider the operator in this space which is represented by the following block matrix:

$$\begin{pmatrix} \mathcal{L}(+M) & 0 \\ 0 & \mathcal{L}(-M) \end{pmatrix} .$$

Furthermore, let us consider the operator which acts on the stochastic variables as the matrix $\begin{pmatrix} -1 & 1 \\ 1 & -1 \end{pmatrix}$. It is represented by the block matrix:

$$\begin{pmatrix} -I_q & I_q \\ I_q & -I_q \end{pmatrix} .$$

The following linear combination

$$\hat{\mathcal{L}}_\varkappa = \begin{pmatrix} \mathcal{L}(+M) & 0 \\ 0 & \mathcal{L}(-M) \end{pmatrix} + i\varkappa \begin{pmatrix} I_q & -I_q \\ -I_q & I_q \end{pmatrix}$$

is the averaged Hamiltonian as the following result shows

Theorem [4]: The quantum evolution operator averaged over the set of all trajectories of the magnetic momentum starting in the stochastic state β at $t = 0$ and ending in the stochastic state α at $t = T$ coincides with the element $\bar{\hat{U}}_{\alpha\beta}(T)$ of the operator matrix which satisfies the following differential equation:

$$\frac{1}{i}\frac{\partial \bar{\hat{U}}}{\partial t} = \hat{\mathcal{L}}_{\varkappa}\bar{\hat{U}}, \qquad \bar{\hat{U}}\Big|_{t=0} = \begin{pmatrix} I_q & 0 \\ 0 & I_q \end{pmatrix}. \tag{10}$$

Together with $\hat{\mathcal{L}}_{\varkappa}$ we shall consider the corresponding unperturbed operator $\hat{\mathcal{L}}^{0}_{\varkappa}$ the quantum part of which is simply the orthogonal sum of operators, acting in the external and internal space, respectively. The external and internal parts of this operator are:

$$-\hat{\Delta} = \begin{pmatrix} -\Delta & 0 \\ 0 & -\Delta \end{pmatrix} + i\varkappa\begin{pmatrix} I_e & -I_e \\ -I_e & I_e \end{pmatrix}, \tag{11}$$

$$\hat{A} = \begin{pmatrix} A^u & 0 \\ 0 & A^d \end{pmatrix} + i\varkappa\begin{pmatrix} I_i & -I_i \\ -I_i & I_i \end{pmatrix}, \tag{12}$$

where $A^u = A(+M)$, $A^d = A(-M)$.

The unperturbed operator $\hat{\mathcal{L}}^{0}_{\varkappa} = (-\hat{\Delta}) \oplus \hat{A}$ is normal and its spectral characteristics can be calculated explicitly. For example, the spectrum of this operator is the sum of the spectrum of $(-\hat{\Delta})$ whose spectrum is purely continuous and consists of two branches: $\lambda = k^2$ and $\lambda = k^2 + 2i$, $\mathrm{Im}\, k = 0$, and of the spectrum of the operator $\hat{A}$ which consists of four simple eigenvalues:

$$\lambda^{\pm}_{0,1}(\hat{A}) = \lambda_{0,1} + i\varkappa \pm \sqrt{M^2 - \varkappa^2}. \tag{13}$$

Operator $\hat{\mathcal{L}}_{\varkappa}$ differs from $\hat{\mathcal{L}}^{0}_{\varkappa}$ by a finite-rank operator only, and this fact allows us to calculate all important spectral characteristics of $\hat{\mathcal{L}}_{\varkappa}$ (see [4]).

The discrete spectrum of the operator $\hat{\mathcal{L}}_{\varkappa}$ consists also of four simple eigenvalues which equal for small coupling parameter α to

$$\lambda^{\pm}_{0,1}(\hat{\mathcal{L}}_{\varkappa}) = \lambda^{\pm}_{0,1}(\hat{A}) \mp \frac{\alpha^2 i}{32\pi} \; \frac{F_{0,1}(\lambda^{+}_{0,1}(\hat{A}))}{(M^2 - \varkappa^2)^{1/2}} + O(\alpha^3), \quad (14)$$

where the functions $F_{0,1}(\lambda)$ are for $\varkappa \ll \lambda_{0,1}$, i.e., for small velocity of the stochastic process equal to

$$F_{0,1}(\lambda) = \frac{2(2\lambda - i\varkappa)}{\sqrt{\lambda}}\Big[(i\varkappa - \lambda)(\lambda^2_{0,1} - M^2 + \lambda_{0,1}(i\varkappa - \lambda) - 1) - \lambda_{0,1}\Big] + O(\varkappa^2) .$$

In Ref. [4] we have calculated the eigenfunctions corresponding to the discrete spectrum of the operator $\hat{\mathcal{L}}_{\varkappa}$:

$$\Psi_k = \begin{pmatrix} v_u & \eta_u \\ v_d & \eta_d \end{pmatrix}_{\lambda_k} \in \mathcal{H} , \qquad k = 1,\ldots,4,$$

where $v_u, v_d \in L^2(\mathbb{R}^3)$, $\eta_u, \eta_d \in \mathbb{C}^2$ and $\lambda_{1,2} = \lambda^{\pm}_0(\hat{\mathcal{L}}_{\varkappa})$, $\lambda_{3,4} = \lambda^{\pm}_1(\hat{\mathcal{L}}_{\varkappa})$. The external part of the eigenfunctions are equal to:

$$\Psi_k(x) = \begin{pmatrix} v_u \\ v_d \end{pmatrix}_{\lambda_k}(x) = \frac{1}{8\pi|x|}\cdot\Big\{C_0 \exp(i\sqrt{\lambda_k}\cdot|x|)\begin{pmatrix} 1 \\ 1 \end{pmatrix} +$$

$$+ C_{\varkappa} \exp(i\sqrt{\lambda_k - 2i\varkappa}\cdot|x|)\begin{pmatrix} 1 \\ -1 \end{pmatrix}, \quad (15)$$

$$0 < \arg\sqrt{\lambda_k} < \pi , \qquad 0 < \arg\sqrt{\lambda_k - 2i\varkappa} < \pi \quad ;$$

$$C_{0,\varkappa} = \text{const}\cdot\alpha(C^-_0 B^+_0 \pm i\varkappa B^-_0) \quad \text{when} \quad \lambda = \lambda_{1,2} ,$$

$$C_{0,\varkappa} = \text{const}\cdot\alpha(C^-_1 B^+_1 \pm i\varkappa B^-_1) \quad \text{when} \quad \lambda = \lambda_{3,4} ,$$

where we use the following notation:

$$A^{\pm}_0 = \lambda_0 \pm M, \quad A^{\pm}_1 = \lambda_1 \mp M, \quad B^{\pm}_{0,1} = A^{\pm}_{0,1} - i,$$
$$C^{\pm}_{0,1} = A^{\pm}_{0,1} + i\varkappa - \lambda, \quad D^{\pm}_{0,1} = (i\varkappa - \lambda)A^{\pm}_{0,1} - 1 . \quad (16)$$

The absolute continuous spectrum (in the sense of [7]) of the operator $\hat{\mathcal{L}}_{\varkappa}$ coincides with the absolute continuous spectrum of unperturbed operator $\hat{\mathcal{L}}^0_\varkappa$, because $\hat{\mathcal{L}}_\varkappa$ differs from it by a finite-rank operator only. Moreover, the operator $\hat{\mathcal{L}}_\varkappa$ has no eigenvalues of infinite multiplicity.

The so-called "scattered waves" play the role of eigenfunctions of $\hat{\mathcal{L}}_\varkappa$. The eigenfunctions in which the initial plane wave is symmetric with respect to the stochastic variables correspond to the $\mathbb{R}^+$ branch of the spectrum (we call it the stable branch). They are of the following form:

$$\Psi_s(x,\nu,\lambda) = \exp(-i\sqrt{\lambda}\cdot\langle x,\nu\rangle)\begin{pmatrix}1\\1\end{pmatrix} + \tag{17}$$

$$+ f_{00}(\lambda)\frac{\exp(i\sqrt{\lambda}\cdot|x|)}{4\pi|x|}\begin{pmatrix}1\\1\end{pmatrix} + f_{10}(\lambda)\frac{\exp(i\sqrt{\lambda-2i\varkappa}\cdot|x|)}{4\pi|x|}\begin{pmatrix}1\\-1\end{pmatrix}.$$

The complex branch $\mathbb{R}^+ + 2i\varkappa$ of the spectrum of $\hat{\mathcal{L}}_\varkappa$ will be called the relaxation branch. It corresponds to the eigenfunctions in which the initial plane wave is antisymmetric with respect to the stochastic variables:

$$\Psi_{as}(x,\nu,\lambda) = \exp(-i\sqrt{\lambda-2i\varkappa}\cdot\langle x,\nu\rangle)\begin{pmatrix}1\\-1\end{pmatrix} + \tag{18}$$

$$+ f_{01}(\lambda)\frac{\exp(i\sqrt{\lambda}\cdot|x|)}{4\pi|x|}\begin{pmatrix}1\\1\end{pmatrix} + f_{11}(\lambda)\frac{\exp(i\sqrt{\lambda-2i\varkappa}\cdot|x|)}{4\pi|x|}\begin{pmatrix}1\\-1\end{pmatrix}.$$

Let us remark that using the methods of the extension theory in $L^2(\mathbb{R}^3)$, we restrict ourselves automatically to the case when the scattering takes place only in the s-channel because the singular solutions in the higher channels are not square integrable. This is why the quantities $f_{i,k}$ in (17), (18) do not depend on the angular variables. The explicit expressions for $f_{i,k}$ were obtained in [4]. For example,

$$f_{00}(\lambda) = \frac{\alpha^2}{4}\sum_{k,m=0}^{1}\left[\frac{\lambda_k - (-1)^{k+m}\cdot M - \lambda}{L(\lambda_k)}\times\right. \tag{19}$$

$$\times\left.\frac{\lambda_k + (-1)^{k+m}\cdot M - i}{L_1(\lambda,\lambda_k)} + \frac{2i\varkappa(\lambda_k - i)}{L(\lambda_k)L_1(\lambda,\lambda_k)}\right]\times Z_k^m$$

where we use the following notation:

$$L(\lambda_k) = (\lambda_k - i)^2 - M^2 ,$$

$$L_1(\lambda,\lambda_k) = (\lambda_k + i\varkappa - \lambda)^2 - M^2 + \varkappa^2 , \tag{20}$$

$$Z_k^0 = Z_k^+ , \quad Z_k^1 = Z_k^- , \quad Z_k^{\pm} = i \; B_k^{\pm}A_k^{\mp} - D_k^{\pm}B_k^{\mp} .$$

3. Eigenfunction-expansion theorem

To formulate the eigenfunction-expansion theorem we have to notice first that the evolution of special elements of quantum-stochastic space whose "internal" components are initially equal to zero is traced only:

$$\hat{f} = \begin{pmatrix} f_u & \zeta_u \\ f_d & \zeta_d \end{pmatrix}, \qquad \zeta_u = \zeta_d = 0. \tag{21}$$

The following assertion is valid

Theorem. Let a vector $\hat{f} \in \mathcal{H}$ have the form (21) and denote its external part as

$$[\hat{f}]_{ext}(x) = \begin{pmatrix} f_u \\ f_d \end{pmatrix}(x) = f(x).$$

Then the following representation for the vector-function $f(x)$ holds almost everywhere (in the Lebesgue measure sense):

$$f(x) = \sum_{n=1}^{4} K_n \int_{\mathbb{R}^3} dy \left\{ \Psi_n(x) \times [\varphi_n(y)]^T \right\} \times f(y) +$$

$$+ \frac{1}{16\pi^3} \int_{\mathbb{R}^+} k^2 dk \int_{\mathbb{R}^3} dy \int_{S^2} d\nu \left\{ \Psi_s(x,\nu,k) \times \left[\varphi_s(y,\nu,-k)\right]^T + \right. \tag{22}$$

$$\left. + \Psi_{as}(x,\nu,k) \times \left[\varphi_{as}(y,\nu,-k)\right]^T \right\} \times f(y).$$

The vector-valued function $\Psi_n(x)$ in the rhs of (22) is the external part of the eigenfunction corresponding to the n-th eigenvalue of $\hat{\mathcal{L}}_{\varkappa}$ (see (15)); the vector-valued $\varphi_n(x)$ is the eigenfunction corresponding to the n-th eigenvalue of $\hat{\mathcal{L}}^*_{\varkappa}$ (this function has a structure analogous to that of $\Psi_n(x)$ and K_n are some constants which can be calculated explicitly. The functions $\Psi_s(x,\nu, k)$ and $\Psi_{as}(x,\nu,k)$ are the eigenfunctions corresponding to different branches of the continuous spectrum of the operator $\hat{\mathcal{L}}_{\varkappa}$ (see the expressions (17) and (18), where $\lambda = k^2$ and $\lambda = k^2 + 2i$ respectively), and the functions $\varphi_s(y,\nu,-k)$, $\varphi_{as}(y,\nu,-k)$ defined by the expressions analogous to (17), (18) are the eigenfunctions corresponding to the continuous spectrum of $\hat{\mathcal{L}}^*_{\varkappa}$.

The proof of this theorem can be carried out by the standard method of integrating the bilinear form of the resolvent of the operator $\hat{\mathcal{L}}_{\varkappa}$ around its spectrum. The present theorem will be called the eigenfunction-expansion theorem.

Using this result in combination with the Fourier method we can construct the external part of the averaged evolution operator. Introducing the notation

$$\left[P_{ext}\,\hat{f}\right](x) = \left[\hat{f}\right]_{ext}(x),$$

we obtain:

$$\left[P_{ext}(\bar{\hat{U}}(t)\hat{f})\right](x) = \left[P_{ext}(\exp(i\,\hat{\mathcal{L}}_{\varkappa}\,t)\hat{f}\right](x) =$$

$$=\sum_{n=1}^{4} K_n \exp(i\lambda_n t)\int_{\mathbb{R}^3} dy \left\{\Psi_n(x)\times\left[\varphi_n(y)\right]^T\right\}\times f(y) +$$

$$+\frac{1}{16\pi^3}\int_{\mathbb{R}^+}\exp(ik^2t)k^2dk\int_{\mathbb{R}^3}dy\int_{S^2}d\nu\left\{\Psi_s(x,\nu,k)\times\left[\varphi_s(y,\nu,-k)\right]^T\right\}\times$$

$$\times f(y) + \frac{1}{16\pi^3}\int_{\mathbb{R}^+}\exp(i(k^2+2i\varkappa)t)k^2dk\int_{\mathbb{R}^3}dy\int_{S^2}d\nu\left\{\Psi_{as}(x,\nu,k)\times\right.$$

$$\left.\times\left[\varphi_{as}(y,\ ,-k)\right]^T\right\}\times f(y)\ . \tag{23}$$

The expression (23) can be used to compute the averaged scattering operator.

4. Scattering operator

In the final part of this paper, we shall describe the result of calculation of the averaged scattering operator for the model under consideration. The averaging of the quantum evolution leads to an evolution-operator semigroup with the generator $\hat{\mathcal{L}}_{\varkappa}$:

$$\bar{\hat{U}}(t) = \exp(i\,\hat{\mathcal{L}}_{\varkappa} t), \qquad t > 0 .$$

which acts in the quantum-stochastic space $\mathcal{H}$. From the formula (23) we see that the contribution of the relaxation branch tends to zero for $t \longrightarrow \infty$. Hence in scattering process only the contribution of the stable branch of spectrum of the operator $\hat{\mathcal{L}}_{\varkappa}$ must be considered. We choose the restriction of the unperturbed operator $(-\hat{\Delta})$ to the stable invariant subspace, corresponding to the real branch $\mathbb{R}^{+}$ of the absolute continuous spectrum of $(-\hat{\Delta})$ as a comparison operator. This leads to the following choice of the identification operator: $J = P_0$, where P_0 is the projector to the subspace consisting of the functions which are symmetric with respect to the stochastic variables. This subspace is invariant under the unperturbed external operator. In this way we eliminate the relaxation branch of the unperturbed-operator spectrum and the scattering matrix looks as follows

$$\bar{S}(\alpha, -\hat{\Delta}_0) = \operatorname*{s-lim}_{t \to +\infty} \exp(i\hat{\Delta}_0 t)\; J\, \exp(2i\,\hat{\mathcal{L}}_{\varkappa} t)\; J^{*} \times \exp(i\hat{\Delta}_0 t) \; . \tag{24}$$

This expression is close to the S-matrix derived in [8]. The only difference is that in the present case the evolution on the complementary subspace of antisymmetric states is not unitary.

The restriction of the operator $(-\hat{\Delta})$ to the symmetric subspace coincides simply with the Laplacian on two-component vectors of the type $\hat{f} = \begin{pmatrix} f \\ f \end{pmatrix}$ with $f \in W_2^2(\mathbb{R}^3)$. Hence we can write $(-\Delta)$ instead of $(-\hat{\Delta}_0)$ in the formula (24) inserting at the same time the projection operator into J, or equivalently, using the operator Σ which averages the stochastic states,

$$\Sigma : \hat{f} = \begin{pmatrix} f \\ f \end{pmatrix} \longmapsto 1/2\,(f + f) = f\ .$$

Also we have to insert in J the operation of the dimension doubling $\begin{pmatrix} 1 \\ 1 \end{pmatrix}$ of the vector f:

$$\begin{pmatrix} 1 \\ 1 \end{pmatrix} : f \longmapsto \hat{f} = \begin{pmatrix} f \\ f \end{pmatrix} .$$

Taking into consideration all these facts, we obtain the averaged scattering operator $L^2(\mathbb{R}^3) \longmapsto L^2(\mathbb{R}^3)$ in the following form:

$$\bar{S}(\alpha,\varkappa) = \operatorname*{s-lim}_{t\to+\infty} \exp(i\Delta t)\, \Sigma\, J\, \bar{\hat{U}}(2t)\, J^{*} \begin{pmatrix} 1 \\ 1 \end{pmatrix} \exp(i\Delta t)\ . \tag{25}$$

Using the expression (23) for the averaged evolution operator, we can obtain the following expression for the averaged scattering matrix in the momentum representation:

$$\left[\hat{\bar{S}}(\alpha,\varkappa)\hat{u}\right](p) = \hat{u}(p) + \frac{i|p|}{2\pi}\cdot f_{00}(p^2)\langle\hat{u}\rangle(|p|)\ . \tag{26}$$

Here $\hat{u} = \mathcal{F}u$ is the Fourier transform of the function from the Schwartz class $\mathcal{S}(\mathbb{R}^3)$, which is dense in $L^2(\mathbb{R}^3)$; $\hat{\bar{S}}(\alpha,\varkappa)$ denotes the scattering matrix in the momentum representation and $\langle\hat{u}\rangle$ is the average of the function $\hat{u}(p) = \hat{u}(|p|\cdot\nu)$ over angular variables,

$$\langle\hat{u}\rangle(|p|) = \frac{1}{4\pi}\int_{S^2} d\nu\ u(|p|\cdot\nu)\ . \tag{27}$$

The kernel of the averaged S-matrix can be written as follows

$$\hat{\bar{S}}(p,p') = \delta(p - p') + \frac{i}{4\pi^2} f_{00}(p^2)\,\delta(p^2 - p'^2), \tag{28}$$

or equivalently,

$$\hat{\bar{S}}(p,p') = \delta(p - p') + \frac{i}{8\pi^2} f_{00}(p^2)\,\frac{\delta(|p| - |p'|)}{|p|}\ . \tag{29}$$

In conclusion, let us consider a very slow Markov stochastic process when the parameter $\varkappa$ tends to zero. Using the explicit expression (19) for the value $f_{00}(p^2)$, we obtain:

$$\lim_{\varkappa \to 0} f_{00}(p^2) = \frac{\alpha^2}{4} \cdot \left\{ \Delta^+(p^2) + \Delta^-(p^2) \right\} , \tag{30}$$

where

$$\Delta^{\pm}(p^2) = \sum_{k=0}^{1} \frac{(\lambda_k \pm (-1)^k \cdot M)p^2 + 1}{\lambda_k \pm (-1)^k \cdot M - p^2} . \tag{31}$$

Thus, the action of the averaged S-matrix on the function $\hat{u}(p)$ is in this case given by the following expression:

$$\hat{\bar{S}}(\alpha, 0) = \hat{u}(p) + \frac{i|p|}{2\pi} \cdot \frac{\alpha^2}{4} \left\{ \Delta^+(p^2) + \Delta^-(p^2) \right\} \langle \hat{u} \rangle (|p|). \tag{32}$$

We can see that the averaged scattering amplitude in this limit coincides with $f_{00}(p^2)$ and equals in the s-channel to

$$f_0 = \lim_{\varkappa \to 0} f_{\varkappa} = \frac{\alpha^2}{4} \left\{ \Delta^+(p^2) + \Delta^-(p^2) \right\} . \tag{33}$$

On the other hand, we can calculate the "deterministic" S-matrices $S^{\pm}$ corresponding to the fixed stochastic states, i.e., the S-matrices corresponding to the operators $\mathcal{L}(+M)$, $\mathcal{L}(-M)$, respectively. They are given by

$$\left[\hat{S}^{\pm}\hat{u} \right](p) = \hat{u}(p) + \frac{i|p|}{2\pi} \cdot \alpha^2 \Delta^{\pm}(p^2) \langle \hat{u} \rangle (|p|) . \tag{34}$$

The corresponding scattering amplitudes are not trivial in the s-channel only and are equal to:

$$f^{\pm} = \alpha^2 \Delta^{\pm}(p^2) . \tag{35}$$

It means that f_0 is equal to $1/4\ (f^+ + f^-)$ and not to $1/2\ (f^+ + f^-)$. It is connected with the fact that calculating the values f^+ and f^- we automatically take into account both branches of the spectrum of the unperturbed operator which coincide for $\varkappa = 0$. In the limit, however, when t tends to infinity for $\varkappa > 0$ the contribution from the complex branch vanishes.

References

1 F.A.Berezin, L.D.Faddeev, DAN SSSR, 137(1961), 1011-1014 (in Russian)

2 S.Albeverio, F.Gestesy, R.Høegh-Krohn, H.Holden : Solvable Models in Quantum Mechanics, Springer, Berlin 1988

3 B.S.Pavlov, Teor. i Matem. Fizika (TMF), 59(1984), 345-353 (in Russian)

4 B.S.Pavlov, A.E.Ryzhkov, Problemy Matem. Fiziki, 1987, iss. 12, 54-82 (in Russian)

5 G.N.Belozerskij, K.A.Makarov, B.S.Pavlov, Leningrad State Univ. Vestnik, 1982, iss. 1, N 4, 12-18 (in Russian)

6 S.E.Cheremshantsev : Some Generalizations of the Feynman-Kac Formula, Leningrad 1982, VINITI N 2195-82 Dep. (in Russian)

7 L.A.Sachnovich, DAN SSSR, 167(1966), 760-763 (in Russian)

8 B.S.Pavlov, Problemy Matem. Fiziki, 1982, iss. 10, 183-208 (in Russian)

FEW-BODY PROBLEMS

FADDEEV EQUATIONS FOR THREE COMPOSITE PARTICLES

Yu. A. Kuperin
Department of Physics, Leningrad State University
Leningrad 198904, USSR

Abstract

A general formulation of the quantum scattering theory for a system of few particles, which have an internal structure, is given. Due to freezing out the internal degrees of freedom in the external channels a certain class of energy-dependent potentials is generated. By means of potential theory modified Faddeev equations are derived both in external and internal channels. We prove the fredholmity of these equations, what provides a sound basis for solving the addressed scattering problem.

1. Introduction

The nonrelativistic quantum mechanics for two and three composite particles has been constructed in different forms and generalized schematically for systems with any number of particles [1 - 12] . Nevertheless, up to now it is rather a heuristic scheme than a mathematically well-defined physical theory, since it is not clear how the basic objects of this formalism, i. e., the energy-dependent potentials, can be treated from the operator point of view. In fact, already in the two-body problem there is no self-adjoint (s.a.) Hamiltonian, which could be generated by a time-dependent unitary group of operators. In the three-body case the original Faddeev equations are also not directly applicable due to the absence of the corresponding s.a. Hamiltonian. There are many other practical and theoretical questions which should be answered before one can seriously applay the formalism to real systems. The critical point of all these questions is the operator interpretation of the theory: how to include the energy-dependent potentials into the few-body Hamiltonians in a mathematically consistent way?

In order to overcome these difficulties we introduce a new class of multichannel few-body scattering models. The channels appearing in these models can be divided into two parts: the so-called external channels describing the standard two - and three- particle scattering, and some additional (internal) channels describing the internal structure of the particles. The Hamiltonians describing the dynamics in the external channels are the usual two - and three - particle Schrödinger operators, while the internal-channel Hamiltonians are given by some abstract operators having only purely discrete spectrum. The coupling of the external and internal channels is in our model realized in the framework of the s.a. extensions theory [13 - 15] leading in such a way automatically to a s.a. Hamiltonian of the global system.

In order to introduce the energy-dependent potentials we proceed further as follows: we exclude the internal channels from the global Hilbert space obtaining in such a way the modified Faddeev equations for the external-channel components of the global Green's function [16, 17] . We prove that these equations are of the Fredholm type. The energy-dependent potentials appear naturally when projecting the global Hamiltonian on the subspace of the external channels. The Fredholm property of the corresponding modified Faddeev equations justifies in some sense the applications to the three-particle scattering problem.

2. Two-body problem

In this section, we construct a s.a. Hamiltonian for two particles possessing a nontrivial internal structure, restricting ourselves to the two-body processes only. We construct also in the external channel the corresponding energy-dependent potential and investigate its properties.

The physical content of the model is fixed by the following assumptions:

(i) In the reaction $X_i + Y_i \longrightarrow C_{if} \longrightarrow X_f + Y_f$, where the indices i, f denote the initial and final channels, respectively, and belong to an index set $\mathscr{C}$, the internal structure of the colliding objects is manifested only at relative distances of the order of the characteristic radius of the particles and not in the asymptotic states.

We separate also the two-body configuration space $\mathbb{R}^3$ (with the centre of mass removed) into two domains $V^{\pm}$, such that $\mathbb{R}^3 = V^- \cup V^+$. Let V^- be the part of the space $\mathbb{R}^3$ where the relative distance between the particles is bounded: $r_i < b_V \sim (\mathrm{Vol}\, V^-)^{\frac{1}{3}} < \infty$. Physically the compact domain V^- can be interpreteted as the reaction domain, and the complement $V^+ = R^3 \setminus V^-$ as the region, where the particles are "asymptotically free". Let us denote as the common boundary of the domains $V^{\pm}$.

We assume that there exist an interaction in the channel $i \in \mathscr{C}$ such that a phase transition occurs at the surface γ which gives rise to the formation of a compound state C_{if} with a finite life-time. At the same time, at $r < b_V$ two types of dynamics are possible: 1) in C_{if} there is an admixture of the "cluster dynamics" of the asymptotic region V^+ ; 2) with a certain probability there is formed in C_{if} a new state u^{in} possessing additional degrees of freedom not contained in the asymptotic region V^+.

(ii) The compound state C_{if} decays into the outgoing channel $f \in \mathscr{C}$, whose energy is equal to the energy of the incoming channel.

From the mathematical point of view, the proposed model is based, as we have already mentioned, on the theory of extensions of symmetric operators. In the two-body case, it is assumed that the global dynamics, which takes into account the interaction of the internal and external degrees of freedom, is specified by a s.a. operator h of a special structure. Namely, it acts on the space $\mathscr{H} = \mathscr{H}^{ex} \oplus \mathscr{H}^{in}$ where $\mathscr{H}^{ex}$ is the Hilbert state space describing the motion of of the particles with the internal degrees of freedom neglected, and $\mathscr{H}^{in}$ is the Hilbert space of the states corresponding to the independent dynamics in the internal degrees of freedom. The method of constructing the operator h in the space $\mathcal{H}$ is the following. Suppose that s.a. operators h^{ex} and A act on the spaces $\mathcal{H}^{ex}$ and $\mathcal{H}^{in}$ respectively. Then the orthogonal sum $h^{ex} \oplus A$ describes the independent dynamics in the external and internal degrees of freedom. An interaction between these dynamics can be "switched in" as follows. We restrict the operators h^{ex} on $\mathcal{H}^{ex}$ and A on $\mathcal{H}^{in}$ to symmetric ones h_0^{ex} and A_0 and construct all s.a. extensions of the operator $h_0^{ex} \oplus A_0$ on the space $\mathcal{H}$. Then each s.a. extension h can be interpretated as a total Hamiltonian defining the coupled dynamics. The nature of the interaction between the external and internal degrees of freedom is regulated both by the method of restricting the operators h^{ex} and A and by the choice of the extension scheme.

We emphasize that in the proposed model the pairs h^{ex}, $\mathcal{H}^{ex}$ and A, $\mathcal{H}^{in}$ can have quite different structures; in particular, we may take for h^{ex}, $\mathcal{H}^{ex}$ a differential operator and an infinite- dimensional function space, respectively, but for $\mathcal{H}^{in}$, A a finite - dimensional space and any s.a. matrix on it. This "freedom" has a lot of advantages, but it leads to a difficulty associated with the fact that the domain of the restricted operator A_0 is not dense. However, this difficulty can be overcome, and moreover, as it is shown below, one can match the internal and external Hamiltonians by means of boundary conditions. The advantage is that one can describe explicitly the S matrix in this way and study in detail its analytical properties. It can be shown that the S matrix inherits the spectral properties and characteristics of the operator A only (in the case of a finite - dimensional matrix A this means the set of its eigenvalues). Finally, the fact that the operators h^{ex} and A can be of a different kind makes it possible to model situations in which the dynamics in $\mathcal{H}^{ex}$ and $\mathcal{H}^{in}$ can differ substantially. An example of such a situation is the hadron-hadron scattering in the bag model [4 - 6] . In hadron-hadron collisions the external dynamics at low and intermediate energies is nonrelativistic and can be described by a Schrödinger equation, whereas within the compound quark bag the dynamics is assumed to be essentially relativistic and given by an orthogonal sum of single - particle Dirac operators.

3. Hamiltonians of the External and Internal Channels.

We follow here the general construction described in [13-17]. Let us assume that the dynamics of the external channel is given by the s.a. Hamiltonian h^{ex} defined as follows

$$h^{ex} u = -(\Delta + v(x))\, u$$

in the Hilbert space $\mathcal{H}^{ex} = L^2(\mathbb{R}^3)$. The potential $v(x)$ represents the so-called peripheral interaction (e.g., a meson - exchange potential) of strongly interacting particles and will be assumed to decrease rapidly as $|x| \to \infty$ and to be sufficiently smooth.

In our model we shall restrict the s.a. Hamiltonian h^{ex} to the symmetric operator h_0 with the domain $D(h_0) = C_0^\infty(R^3 \setminus \gamma)$, where $C_0^\infty(R^3 \setminus \gamma)$ is the class of infinitely differentiable functions, which vanish together with all derivatives in the neighbourhood of the surface γ. The adjoint operator h_0 has a nontrivial boundary form $J^{ex}(\cdot,\cdot)$, namely

$$J^{ex}(u,f) = \langle h_0^* u, f\rangle - \langle u, h_0^* f\rangle = \tag{1}$$

$$= \lim_{\delta \to 0} \Big(\int_{\gamma_\delta^+} dS\, (n\cdot\nabla u \bar{f} - u\, n\cdot\nabla \bar{f}) -$$

$$- \int_{\gamma_\delta^-} dS\, (n\cdot\nabla u \bar{f} - u\, n\cdot\nabla \bar{f})\Big),$$

where $n\cdot\nabla_{\pm}$ is the normal derivative on the surfaces $\gamma_\delta^{\pm} = \{ x \in V^{\pm} : \text{dist}(x, \gamma^{\pm}) = \delta \}$.

Now we asuume that the dynamics of the internal degrees of freedom without a coupling to the external channel $\mathcal{H}^{ex}$ is given by an arbitrary s.a. operator A acting in some Hilbert space $\mathcal{H}^{in}$. The important question of the model is the following: how to parametrize the boundary form $J^{in}(\cdot,\cdot)$ for an arbitrary symmetric operator A_0 ?

The general answer was obtained in Ref. [13] . For the symmetric operator A_0 with a non-dense domain $D(A_0)$ this question was also solved by B. S. Pavlov (see his contribution in these proceedings). Following the general construction, the symmetric restriction A_0 of the s.a. Hamiltonian A should be performed in our scheme via its Cayley transform $U = (A - iI)(A + iI)^{-1}$. Let us consider the special isometric restriction $U_0 = U \restriction (U^* \theta)^{\perp}$, where θ is a generating element of the operator A . The symmetric restriction A_0 can be obtained as the inverse Cayley transform of the isometry U_0 . Hence the operator A_0 has deficiency indices (1,1) and the domain $D(A_0^*)$ of its adjoint can be described in terms of von Neumann theory : $D(A_0^*) = D(\bar{A}_0) \dot{+} \mathcal{L}(\theta, U^*\theta)$. Here $\bar{A}_0$ is the closure of the A_0 and $\mathcal{L} := \mathcal{L}(\theta, U^*\theta)$ is the span of deficiency elements θ and $U^*\theta$. Let us emphasize that though von Neumann formulas cannot be directly used in the case of a non-densely defined operator A_0 the description they give for the domain of A_0 remains valid.

It is convenient to introduce some new basis in $\mathcal{L}$: $w^+ = (1/2)(U^*\theta + \theta)$, $w^- = (1/2i)(U^*\theta - \theta)$. In accordance with the first von Neumann's theorem an arbitrary element $u \in D(A_0^*)$ can be decomposed as

$$u = \tilde{u} + \varepsilon^+ w^+ + \varepsilon^- w^- , \quad u \in D(\bar{A}_0) ,$$

where $\varepsilon^{\pm}(u)_{\pm}$ are the so-called boundary values of the element u . In terms of $\varepsilon_{\pm}^{\pm}$, the boundary form of the operator A_0^* reads $A_0^* w^{\pm} = \mp w^{\mp}$, and it can be written as the symplectic form in the boundary-value space [13, 15] :

$$J^{in}(u_1,f_1) = \langle A_0^* u_1,f_1\rangle - \langle u_1, A_0^* f_1\rangle = \tag{2}$$

$$= \varepsilon^-(u_1)\cdot\overline{\varepsilon^+(f_1)} - \varepsilon^+(u_1)\cdot\overline{\varepsilon^-(f_1)} , \qquad u_1, f_1 \in \mathcal{H}^{in} .$$

After we have constructed the boundary forms $J^{ex,in}(.,.)$, the next step is to construct a s.a. extension h of the operator $h_0 \oplus A_0$, acting in the orthogonal sum $\mathcal{H}^{ex} \oplus \mathcal{H}^{in}$. In accordance with our general method one should find such boundary conditions for which the total boundary form vanishes, i.e., for which

$J^{ex}(.,.) + J^{in}(.,.) = 0$. It can be shown that such conditions can be written , for instance, as

$$[n\cdot\nabla u_0]_\gamma = - \varepsilon^-(u_1)\varphi \ , \tag{3}$$

$$\varepsilon^+(u_1) = \langle u_0, \varphi\rangle := \int_\gamma dS\, u_0 \overline{\varphi} \ . \tag{4}$$

Here u_α , $\alpha = 0,1$, are the external and internal channel elements, respectively. We use the notation $[n\cdot\nabla u]_\gamma := n\cdot\nabla u^- - n\cdot\nabla u^+$ to denote the jump of a function $n\cdot\nabla u$ across γ . Here the superscripts $-$, $+$ denote the limits taken from V^- and V^+, respectively. The function $\varphi \in L^2(\gamma)$ is the parameter of the model.

The investigation of the wave functions $\mathcal{U} = (u_0, u_1)$ is based on the two-channel Schrödinger equation

$$(h - z)\mathcal{U} = 0 , \qquad x \in \mathbb{R}^3 \setminus \gamma \tag{5}$$

with the boundary conditions (3), (4) . In order to obtain the energy-dependent potential we consider further only the external channel $\mathcal{H}^{ex}$. For this purpose we have to exclude the internal boundary values $\varepsilon^\pm$ from (3) and (4) . This can be done using the the following relation [13] :

$$\varepsilon^- = \Delta(z)\ \varepsilon^+, \tag{6}$$

where $\Delta(z)$ is the Schwartz integral of the spectral measure $d\langle E_z^A\theta, \theta\rangle$ of the s.a. operator A ,

$$\Delta(z) := \langle (I + zA)(A - zI)^{-1}\theta , \theta\rangle = \tag{7}$$

$$= \int_{\mathbb{R}} (1 + tz)(t - z)^{-1}\, d\langle E_t^A\theta, \theta\rangle .$$

Taking into account the relation (6), we obtain from (3), (4) the following energy-dependent boundary conditions in the external space $\mathcal{H}^{ex}$:

$$[n\cdot\nabla u_0]_\gamma = - \Delta(z)\langle u_0, \varphi\rangle\varphi . \tag{8}$$

In accordance with (5) the external component u_o obeys the equation

$$(h_o^* - z) u_o = 0 , \qquad x \in \mathbb{R}^3 \setminus \gamma . \tag{9}$$

In order to obtain the differential equation in the external configuration space $\mathbb{R}^3$ it is convenient to use a quasipotential approach (see, e.g., [19, 20]). Let us consider the quasipotential $V(z)$ acting on the function u as

$$V(z) u := - \delta_\gamma \Delta(z) \langle u, \varphi \rangle \varphi . \tag{10}$$

Here $\delta_\gamma \mu$ is the distribution, usually called the simple layer [21], that acts on the set of sufficiently smooth functions f in the following way

$$\langle \delta_\gamma \mu , f \rangle := \int_\gamma dS \, \mu \bar{f} .$$

The boundary-value problem in the external space $\mathcal{H}^{ex}$ can be written in terms of quasipotential $V(z)$ as

$$(h^{ex} + V(z) - z) u_o = 0 , \tag{11}$$

where the variable x now runs over the whole configuration space $\mathbb{R}^3$. One can state that (9) is equivalent to the boundary - value problem (8), (9).

We conclude this section with the following remarks:

1. The operator h , which is the total Hamiltonian in the two-body system, is a self-adjoint operator and hence the scattering for the pair ($h, h^{ex} \oplus A$) can be investigated in a mathematically correct way. It should also be noted that in our model we are able to simulate an arbitrary complicated internal structure of particles due to the general nature of the internal self-adjoint operator A . The physical nature of the Hamiltonian A can be interpreted in different ways. In the non-relativistic framework it might be, for example, a few-body Schrödinger operator with confining potentials [18, 19] .

2. As it follows from (7) and (10) the energy dependence of the potentials cannot be arbitrary. It is given by the Schwartz integral $\Delta(z)$, which is real on the real axis and is an analytic function in the upper half-plane $\operatorname{Im} z > 0$ with the positive imaginary part $\operatorname{Im} \Delta(z) > 0$. It can be shown that such interactions ensure the analyticity and unitarity of the corresponding scattering matrix [14, 16].

3. In the model, described above, the quasipotential $W(z)$ is separable and of rank one. The generalization to an arbitrary rank of $W(z)$ is trivial. For this purpose one should increase the dimension of the deficiency subspaces $\mathcal{N} = \{\theta\}$, $\mathcal{N}^* = \{U^*\theta\}$ and change in a self-consistent way the functionals φ, $\langle \cdot, \varphi \rangle$ by arbitrary finite-rank operators B, B^*.

4. The Three-body Problem.

We consider in this section a system of three particles having a nontrivial internal structure. A total s.a. Hamiltonian H governing the dynamics of external and internal degrees of freedom is an most important object in the three-body analysis.

In order to describe the kinematics of the system we use the usual Jacobi coordinates x_α, y_α, $\alpha = 1,2,3$, which we combine into the six-component vector $X = \{x_\alpha, y_\alpha\}$ belonging to the external configuration space $\mathbb{R}^6$ (with the centre of mass removed).

Let $\Gamma_\alpha = \gamma_\alpha \times \mathbb{R}^3_y$ be cylinders in $\mathbb{R}^6$ and $\Gamma = \bigcup_\alpha \Gamma_\alpha$. The starting point of the method is the description of the two-body Hamiltonian

$$H_\alpha := h_\alpha \otimes I_y + I_\alpha \otimes (- \Delta_{y_\alpha}) \tag{12}$$

in the three-body configuration space $\mathbb{R}^6$. Here h_α is the s.a. two-body Hamiltonian defined in the previous section, I_y and I_α are the identity operators in the spaces $L^2(\mathbb{R}^3_y)$ and $\mathcal{H}_\alpha = \mathcal{H}^{ex}_\alpha + \mathcal{H}^{in}_\alpha$, respectively, and $-\Delta_{y_\alpha}$ is the Laplacian defined on its natural domain $W_2^2(\mathbb{R}_y^3)$. The operator H_α is essentially self-adjoint on the domain

$$D(H_\alpha) := D(h_\alpha) \otimes W_2^2(\mathbb{R}_y^3) \tag{13}$$

The closure $\bar{H}_\alpha$ of the operator H_α is the s.a. operator, which will be denoted by the same symbol H_α.

The domain $D(H_\alpha)$ can be also described in terms of boundary conditions. Namely, let $\mathcal{U} = (u_0, u_\alpha) \in D(H_\alpha)$. Then the external component u_0 is a W_2^2 - smooth function outside of Γ_α. The internal component $u_\alpha \in \mathcal{G}_\alpha^{in} := \mathcal{H}_\alpha^{in} \otimes L^2(\mathbb{R}_y^3)$ can be decomposed into the sum

$$u_\alpha = \tilde{u}_\alpha + \varepsilon_\alpha^+(y_\alpha)\, w_\alpha^+ + \varepsilon_\alpha^-(y_\alpha)\, w_\alpha^- \;, \quad \tilde{u}_\alpha \in D(H_{\alpha 0}^{in}) \;, \tag{14}$$

where $w^\pm$ are the deficiency elements of the symmetric operator $A_{\alpha 0}$, which is the restriction of the s.a. operator A_α, and

$$H_{\alpha 0}^{in} := A_{\alpha 0} \otimes I_y + I_\alpha \otimes (-\Delta_{y_\alpha}) \,. \tag{15}$$

Let us note that the formula (14) is a simple consequence of the special structure of the space $\mathcal{G}_\alpha^{in}$.

The functions $\mathcal{U} \in D(H_\alpha)$ satisfy the boundary conditions

$$[n \cdot \nabla u_0]_{\Gamma_\alpha} = -\, \varepsilon_\alpha^-(y_\alpha)\, \varphi_\alpha(x_\alpha), \tag{16}$$

$$\varepsilon_\alpha^+(y_\alpha) = \langle u_0, \varphi_\alpha \rangle (y_\alpha) := \int_{\gamma_\alpha} dx_\alpha\, u_0(X)\, \overline{\varphi_\alpha(x_\alpha)}. \tag{17}$$

It should be noted that the boundary conditions (16), (17) have a two-body character (see the previous section). The only difference is that the $\varepsilon_\alpha^\pm(y_\alpha)$ are now functions of the variable $y_\alpha \in \mathbb{R}_y^3$.

We are now ready to construct the total three-body Hamiltonian H. Let us consider the space

$$\mathcal{G} := L^2(\mathbb{R}^6) \oplus \Big(\sum_{\alpha=1}^{3} \oplus\, \mathcal{G}_\alpha^{in} \Big) \tag{18}$$

and symmetric operator H_0 in $\mathcal{H}$,

$$H_0 \mathcal{U} := \begin{cases} (-\Delta_X + \sum_\alpha v_\alpha(x_\alpha))\, u_0 , \\ H^{in}_{\alpha 0}\, u_\alpha , \quad \alpha = 1,2,3, \end{cases} \tag{19}$$

with the domain

$$D(H_0) := C_0^\infty (\mathbb{R}^6 \setminus \Gamma) \oplus \sum_\alpha \oplus D(H^{in}_{\alpha 0}) \tag{20}$$

Any s.a. extension H of the operator H_0 is a total three-body Hamiltonian describing the whole dynamics in both external and internal channels. In accordance with the von Neumann theory all such extensions can be obtained by the extension of the operator H_0 on its deficiency subspaces. So we shall extend the domain $D(H_0)$ to a linear set $D(\tilde{H})$ in the following way

$$D(\tilde{H}) = \begin{cases} u_0 = \tilde{u}_0 + \sum_\alpha R_0(-1)\rho_\alpha , \quad \tilde{u}_0 \in C_0^\infty(\mathbb{R}^6 \setminus \Gamma) & (21) \\ u_\alpha = \tilde{u}_\alpha + \varepsilon^+_\alpha w^+_\alpha + \varepsilon^-_\alpha w^-_\alpha , \quad \varepsilon^\pm_\alpha \in W_2^2(\mathbb{R}^3_y) . & (22) \end{cases}$$

and impose on the $D(\tilde{H})$ the boundary conditions (16), (17). Here $R_0(z) = (H^{ex} - z)^{-1}$ is the resolvent of the s.a. operator $H^{ex} = -\Delta_X + \sum_\alpha v_\alpha(x_\alpha)$ and ρ_α are densities of the simple-layer potentials given on the cylinders Γ_α , $\alpha = 1,2,3$. The corresponding extension of the operator H_0 on the domain $D(\tilde{H})$ with the boundary conditions (16), (17) will be denoted $\tilde{H}$. In terms of the densities ρ_α these conditions can be written as

$$\rho_\alpha(X) = -\varepsilon^-_\alpha(y)\, \varphi_\alpha(x_\alpha), \tag{23}$$

$$\varepsilon^+_\alpha(y_\alpha) = \langle \sum_{\beta=1}^{3} R_0(-1)\rho_\beta , \varphi_\alpha \rangle (y_\alpha). \tag{24}$$

Let us now state some important facts about the operator $\tilde{H}$.

<u>Theorem 1.</u> The domain $D(\tilde{H})$ is dense in the space $\mathcal{H}$ and the operator $\tilde{H}$ given by $H_0 \restriction D(\tilde{H})$ is symmetric and bounded from below.

The proof of this statement will be given elsewhere.

The last step is now to construct a self-adjoint extension H of the symmetric operator $\tilde{H}$ obeying the following conditions:

1. The translation-invariant boundary conditions (16), (17) hold on D(H).

2. The Hamiltonian H is bounded from below.

For this purpose we shall choose the Friedrichs extension [23] H of the operator $\tilde{H}$. On the domain D(H) which can be described as usual [23], the action of H is given by

$$H\mathcal{U} = \begin{cases} H^{ex}\, u_o , \\ -\Delta_{y_\alpha} u_\alpha + A_\alpha \tilde{u}_\alpha - \varepsilon_\alpha^+ w_\alpha^- + \varepsilon_\alpha^- w_\alpha^+ \end{cases} \qquad (25)$$

$$\mathcal{U} = (u_o, u_\alpha), \quad \alpha = 1,2,3,$$

with the boundary conditions (16), (17).

5. Resolvent Equations.

This section deals with the Fredholm-type equations for the resolvent R(z) of the s.a. Hamiltonian H. As in the case of the energy-independent interactions [22, 24], these equations represent a starting point for the three-body scattering problem.

First we shall derive the differential equations for the resolvent components $R_{ab}(z)$ corresponding to the decomposition of $\mathcal{G}$ into the sum (18),

$$R(z) = \{R_{ab}(z)\}, \quad a, b = 0,1,2,3. \qquad (26)$$

Here the indices a, b stand for the external (a,b = 0) and internal (a,b = 1,2,3) subspaces $\mathcal{G}_0 = L^2(\mathbb{R}^6)$ and $\mathcal{G}_\alpha^{in}$, $\alpha = 1,2,3$, respectively.

Because R(z) is the resolvent of the s.a. operator H it satisfies the usual relations

$$R_{ab}^*(z) = R_{ba}(\bar{z}) \quad ; \quad a,b = 0,1,2,3. \tag{27}$$

We shall introduce the following notations: Let F be an arbitrary element of $\mathcal{G}$ and $\mathcal{U} = R(z)\, F$, i.e., $F = (f_0, f_1, f_2, f_3)$ and

$$u_a = \sum_{b=0}^{3} R_{ab}(z)\, f_b \,. \tag{28}$$

Then due to (25) and (28) one gets

$$\varepsilon_\alpha^{\pm} = \sum_{b=0}^{3} \mathcal{E}_{\alpha b}^{\pm}(z) \cdot f_b \,, \quad \alpha = 1,2,3, \tag{29}$$

where $\mathcal{E}^{\pm}_{\ b}$ are the operators which mapping $\mathcal{G}_0$ into $L^2(\mathbb{R}_y^3)$ for $b = 0$, and $\mathcal{G}_\alpha^{in}$ into $L^2(\mathbb{R}_y^3)$ for $b \neq 0$. The relation (29) can be considered as the definition of these operators.

Let $\tilde{R}_{\alpha b}(z)$ denote the operators

$$\tilde{R}_{\alpha b}\, f_b = (R_{\ b} - w_\alpha^+ \,\mathcal{E}_{\alpha b}^+ - w_\alpha^- \,\mathcal{E}_{\alpha b}^-)\, f_b \,. \tag{30}$$

Then using the identity

$$(H - z)\, R(z)\, F = F$$

one can obtain a set of equations for the kernels of the operators $R_{ab}(z)$ and $\mathcal{E}_{\alpha b}^{\pm}(z)$:

$$(H^{ex} - z)\, R_{0b}(z) = \delta_{0b}\, I_0 \,, \tag{31}$$

$$A_\alpha \tilde{R}_{\alpha b} - w_\alpha^- \mathcal{E}_{\alpha b}^+ + w_\alpha^+ \mathcal{E}_{\alpha b}^- - (\Delta_y + z) R_{\alpha b} = \delta_{\alpha b} I_\alpha , \qquad (32)$$

with the following boundary conditions

$$[n \cdot \nabla R_{0b}]_{\Gamma_\alpha} = - \varphi_\alpha \mathcal{E}_{\alpha b}^- , \qquad (33)$$

$$\mathcal{E}_{\alpha b}^+ \cdot = \langle R_{0b} \cdot , \varphi_\alpha \rangle \restriction \Gamma_\alpha \qquad (34)$$

The differential equations (31), (32) for the external R_{0b} and internal $R_{\alpha b}$ components of the resolvent $R(z)$ serve as a starting point for construction of the Faddeev equations.

We shall rewrite the conditions (33) and (34) in terms of the internal Hamiltonians A_α. For this purpose we use the relation

$$\mathcal{E}_{\alpha b}^- \cdot = Q_\alpha(z) \ \mathcal{E}_{\alpha b}^+ \cdot + \delta_{\alpha b} \langle (A - iI) (H_\alpha^{in} - z)^{-1} \cdot , \theta_\alpha \rangle \qquad (35)$$

which can be obtained by arguments analogous to the two-body case [13, 15]. Here

$$H_\alpha^{in} = A_\alpha \otimes I_y + I_\alpha \otimes (- \Delta_{y_\alpha}) \qquad (36)$$

and $Q_\alpha(z)$ is the generalization of the Schwartz integral in the three-body configuration space

$$Q_\alpha (z) = \langle (I + (\Delta_{y_\alpha} + z) A_\alpha) (H_\alpha^{in} - z)^{-1} \theta_\alpha , \theta_\alpha \rangle . \qquad (37)$$

In accordance with (15) this operator is an integral operator with the kernel

$$Q_\alpha (y_\alpha - y_\alpha' , z) = \frac{1}{2\pi i} \oint_{L_\alpha} dt \ \Delta_\alpha (t) \ r_0^\alpha (y_\alpha - y_\alpha' , z - t). \qquad (38)$$

Here $r_0^\alpha(z) = (- \Delta_{y_\alpha} - z)^{-1}$ is the resolvent of the Laplacian, $\Delta_\alpha (t)$ is the two-body Schwartz integral, and the contour L_α encircles the spectrum of A_α.

The operators $\mathcal{E}_{\alpha b}^\pm$ can now be excluded from (33) and (34) by virtue of the relation (35) :

$$[\, n\cdot\nabla R_{0b}]\Big|_{\Gamma_\alpha} \cdot = -\varphi_\alpha \;(\; Q_\alpha(z) \,\langle\, R_{0b}\cdot \,, \varphi_\alpha\rangle \; + \qquad (39)$$

$$+\,\delta_{\alpha b} \,\langle\, (A \;-\, iI)\, (H^{in} - z)^{-1} \cdot \;,\; \theta_\alpha \,).$$

If the internal-channel Hamiltonians A_α have the discrete spectra $\sigma_p(A_\alpha) = \{\lambda_s^\alpha\}$ only, than the kernels $Q_\alpha(y_\alpha - y_\alpha', z)$ should be written in the form

$$Q_\alpha(y_\alpha - y_\alpha', z) = \sum_s (1 + (\lambda_s^\alpha)^2) \,\langle E_s^A\, \theta_\alpha \,, \theta_\alpha\rangle \times \qquad (40)$$

$$\times\; r_0^\alpha(y_\alpha - y_\alpha' \,, z - \lambda_s^\alpha) \,,$$

where E_s^A are the spectral projectors of the operators A_α .

Notice that internal Hamiltonians are used for describing the internal channels, e.g., with quark confinement.

6. The External-Channel Faddeev Equations.

The study of the total resolvent $R(z)$ can be reduced to constructing the external-channel component $R_{00}(z)$ only. In order to see this, the eqs. (31) - (37) should be used. Let the component $R_{00}(z)$ be given. From (34) we can get $\mathcal{E}_{\alpha 0}^+(z)$ which we use for substitution into (35) to get $\mathcal{E}_{\alpha 0}^-(z)$. Then from (32) and from the definition (30) of the $\widetilde{R}_{\alpha b}(z)$, $\mathrm{Im}\, z \neq 0$, i.e.,

$$R_{\alpha b} = \widetilde{R}_{\alpha b} + w_\alpha^+ \,\mathcal{E}_{\alpha b}^+ + w_\alpha^- \,\mathcal{E}_{\alpha b}^- \,,$$

one can obtain $\widetilde{R}_{\alpha 0}(z)$. It gives the components $R_{\alpha 0}(z)$, $\alpha = 1,2,3$. Then the components $R_{0\alpha}(z)$, $\alpha = 1,2,3$, can be found from (27). Thus we shall now treat further on the component $R_{00}(z)$ only; for the simplicity, it will be denoted by $G(z)$. It should be noted that $G(z)$ is the so-called Krein's resolvent [25, 15] and it has the corresponding properties.

In view of eq. (39) the kernel $G(X, X', z)$ of the quasiresolvent $G(z)$ obeys the boundary conditions

$$[\, n \cdot \nabla G(X,X',z)\,]_{\Gamma_\alpha} \cdot = - \varphi_\alpha(x_\alpha)\, Q_\alpha(z) \langle G(z)\cdot , \varphi_\alpha \rangle . \qquad (41)$$

As in the two-body case these conditions can be written in terms of quasipotentials

$$\mathbb{V}_\alpha(z)\, \mathcal{U} := \delta_{\Gamma_\alpha} V_\alpha(z)\, \mathcal{U} , \qquad (42)$$

where $V_\alpha(z)$ is the integral operator in $L^2(\Gamma_\alpha)$ with the kernel

$$V_\alpha(X,X',z) = - \varphi_\alpha(x\;)\, Q_\alpha(y_\alpha - y'_\alpha , z)\, \overline{\varphi_\alpha(x'_\alpha)} . \qquad (43)$$

In accordance with (31), (41) and (42) we obtain the following equation

$$(H^{ex} + \sum_{\alpha=1}^{3} \mathbb{V}_\alpha(z) - z)\, G(X,X',z) = \delta(X - X'). \qquad (44)$$

To derive an integral equation for Krein's quasiresolvent one can use the usual procedure. Namely, applying the operator $R_0(z) = (H^{ex} - z)^{-1}$ to (44) we obtain the resolvent identity for $G(z)$:

$$G(z) = R_0(z) - R_0(z) \sum_{\alpha=1}^{3} \mathbb{V}_\alpha(z)\, G(z) . \qquad (45)$$

From this equation which is of the Lippman-Schwinger type the operator $G(z)$ can be expressed explicitly in terms of generalized operators

$$M_\alpha(z) = \mathbb{V}_\alpha(z)\, G(z) \qquad (46)$$

by the relation

$$G(z) = R_0(z) - R_0(z) \sum_\alpha M_\alpha(z) . \qquad (47)$$

Hence we have reduced the problem of investigating the quasiresolvent $G(z)$ to study of the operators $M_\alpha(z)$ [20].

The next problem is to derive the Faddeev equations from eq. (47). Applying the operators $V_\alpha(z)$ to (47) one can write this equation in the form

$$(I + V_\alpha R_0) M_\alpha = V_\alpha R_0 - V_\alpha R_0 \sum_{\beta \neq \alpha} M_\beta \quad . \tag{48}$$

Following Faddeev's method we have to invert the operator $I + V_\alpha R_0$. This inversion can be performed explicitly in terms of the two-body operator $G_\alpha(z) = (H_\alpha - z)^{-1}$ which is the resolvent of the s.a. operator H_α. The following formula can be easily verified

$$(I + V_\alpha R_0) V_\alpha G_\alpha = V_\alpha R_0 \quad . \tag{49}$$

This relation yields in a straightforward way the equations

$$M_\alpha(z) = V_\alpha G_\alpha(z) - V_\alpha G_\alpha(z) \sum_{\beta \neq \alpha} M_\beta(z) \tag{50}$$

which have the structure of the Faddeev equations.

Nevertheless to ensure that these equations are actually Faddeev equations, one must prove the following statement:

<u>Theorem 2.</u> Let μ_α be the densities of the simple-layer potentials, $M_\alpha(z) := \delta_{\Gamma_\alpha} \mu_\alpha(z)$ and $\mu = (\mu_1, \mu_2, \mu_3)$. Then:

1. Equations (50) rewritten in terms of densities μ_α,

$$\mu(z) = \mu_0(z) + B(z)\, \mu(z) \ , \tag{51}$$

are of Fredholm type and B^n , $n > N_{max}$, with a sufficiently large N_{max} , is a compact operator in an appropriate Banach space.

2. Equations (50) or (51) are spectral-equivalent to the original Schrödinger equation with the s.a. Hamiltonian H .

The proof of the first statement proceeds in a standard way [22] .

The second statement of the theorem is much more subtle in contrast to the case of energy-independent interactions. In particular, we must show that the homogeneous equations

$$\mu(z) = B(z)\,\mu(z) \tag{52}$$

have a nontrivial solution, iff $z \in \sigma_p(H)$, where $\sigma_p(H)$ is the discrete spectrum of the s.a. operator H.

Let μ be the solution of the homogeneous equations (52) rewritten in the form

$$\mu_\alpha(z) = -V_\alpha(z)\,G_\alpha(z) \sum_{\beta\neq\alpha} \mu_\beta(z)\ . \tag{53}$$

Consider the function

$$u_0 = R_0(z) \sum_\beta \mu_\beta \quad ,$$

which is evidently the simple-layer potential given on the hypersurface $\Gamma = \bigcup_\alpha \Gamma_\alpha$ and hence it satifies the equation

$$(H^{ex} - z)\,u_0(X) = 0\ , \qquad X \notin \Gamma\ . \tag{54}$$

In order to find the appropriate boundary conditions one must apply the operator $I + V_\alpha R_0$ to eq. (53). Taking into account (49) and the properties [21] of the simple-layer potential : $[n\cdot\nabla u_0]_{\Gamma_\alpha} = -\mu_\alpha$, we find the boundary conditions

$$[n\cdot\nabla u_0]_{\Gamma_\alpha} = V_\alpha(z)\,u_0\ . \tag{55}$$

Iterating the eq. (53) one finds that $u_0 \in W_2^2(\mathbb{R}^6 \setminus \Gamma)$ at $\mathrm{Im}\, z \neq 0$ as well at $z = E \pm i0$, $E \in R$, and furthermore, that $\langle u_0\,, \varphi_\alpha\rangle \in W_2^2(R_y^3)$.

Now we shall express the internal functions u_α in terms of the external component u_0 . To this end one must take into account the representation u_α in the form (22) and the relations (23), (24) as well as (35), which express the relation between $\varepsilon_\alpha^\pm$ and u_0 :

$$\varepsilon_\alpha^+ = \langle u_0\,, \varphi_\alpha\rangle \quad , \tag{56}$$

$$\varepsilon_\alpha^- = Q_\alpha(z)\, \varepsilon_\alpha^+ \ . \tag{57}$$

The functions $\tilde{u}_\alpha$ may be then found as solutions of the equations

$$(- \Delta_{y_\alpha} + A_\alpha - z)\, \tilde{u}_\alpha = \varepsilon_\alpha^+(y_\alpha)\, w_\alpha^- - \varepsilon_\alpha^-(y_\alpha)\, w_\alpha^+ + \tag{58}$$

$$+ (\Delta_{y_\alpha} + z)\, (\varepsilon_\alpha^- w_\alpha^- + \varepsilon_\alpha^+ w_\alpha^+) \ .$$

By virtue of equation (56) the functions $\varepsilon_\alpha^+ \in W_2^2(\mathbb{R}_y^3)$, and hence $\varepsilon_\alpha^- \in W_2^2(\mathbb{R}_y^3)$. This means that $\mathcal{U} = (u_0 , u_\alpha)$ belongs to D(H) and, in accordance with (54) - (58), $\mathcal{U}$ is an eigenvector of the s.a. operator H,

$$(H - z)\, \mathcal{U} = 0 \ . \tag{59}$$

This equation implies that $\mathcal{U} = 0$ if $z \notin \sigma_p(H)$ and hence $u_0 = 0$ in this case. In other words we have proven that eq. (51) has a unique solution, if $z \notin \sigma_p(H)$.

On the contrary let $\mathcal{U}$ be an eigenvector of the Hamiltonian H. Then one must repeat the derivation of (50) for densities $\mu_\alpha = - V_\alpha u_0$, which obey the equations (53).

Hence we have proven that the Faddeev equations (53) are spectral-equivalent to the Schrödinger equation (59).

Consequently, the Fredholm alternative can be applied to (51) and the properties of densities μ_α can be investigated. Knowing these properties we can study the behaviour of the resolvent R(z) by (47). The wave functions are determined too. Their asymptotic form can be investigated and their completeness in the total space established using the methods of Ref. [22].

7. Discussion.

In this paper we have presented a new approach towards a mathematically correct study of the scattering theory for few-body systems with energy-dependent potentials. The main result is that treating of such systems in usual configuration space is inconsistent from an operator point of view. We have demonstrated that the energy dependence of the potentials is generated by the internal structure of the interacting particles. This very dependence, however, turns out not be arbitrary, since it is given by some class of operator-valued R-functions, including in particular Schwartz integrals as described above.

The main effect incorporated in our scheme is the possibility to separate the contributions from two-body and three-body forces. From the geometrical point of view the three-body forces are connected with boundary conditions, which may be stated on the manifold $\Gamma_0 = \bigcap_\alpha \Gamma_\alpha$. The deficiency subspaces of the operator H_0 , corresponding to the manifold Γ_0 , are parametrized by simple-layer densities belonging to the Sobolev class $W_2^{-3/2}$. In order to conserve the pair character of the boundary conditions (16), (17) we do not include such deficiency elements into the domain $D(\tilde{H})$. But it is clear that three-body forces can be included into our consideration without a drastic change of the formulation.

Acknowledgments.

I would like to thank my collegues K. A. Makarov, S. P. Merkuriev, A. K. Motovilov and B. S. Pavlov, who collaborated with me on the problems discussed in this paper. I am also indebted to L. D. Faddeev for encouragement and fruitful discussions and to P. Exner and P. Šeba for some essential remarks, which improved the content of the manuscript.

References

1. Feshbach H. Ann. Phys. 5, 357(1958); ibid. 19, 287(1962).
2. Schmid E.W. Proc. of the Intern. Sympos. on Few Particle Problems in Nuclear Physics, Dubna, 1979, p.174.
3. Wildermuth K., Tang Y.C. A Unified Theory of Nucleus, N.Y., Academic Press , 1977.

4. Hill D.A., Wheeler J.A. Phys. Rev. 89, 1102(1953).
5. Alt E.O., Grassberger P., Sandhas W. Nucl. Phys. B2, 167(1967).
6. Jaffe R.L., Low F.E. Phys. Rev. D19, 2105(1979).
7. Simonov Yu.A. Phys. Lett. B107, 1(1981).
8. Narodetskii I.M. : Few-Body Problems in Physics (Ed. by Faddeev L.D. and Kopaleishvili T.I.), Singapore, World Scientific, 1985.
9. Vanzani V., Cattapan G. Phys. Rev. C19, 1168(1979).
10. McKellar B.H.J., McKay C.M. Austr. J. Phys. 36, 607(1983).
11. Orlowski M. Helv. Acta Phys. 56, 1053(1983).
12. McTavish J.P. J. Phys. G8, 1047(1982).
13. Pavlov B.S. Teor. Mat. Fiz. 59, 345(1984).
14. Adamjan V.M., Pavlov B.S. Zap. Nauch. Semin. LOMI, 149, 7(1986).
15. Pavlov B.S. Uspekhi Mat. Nauk 42, 99(1987).
16. Kuperin Yu.A., Makarov K.A., Pavlov B.S. Teor. Mat. Fiz. 63, 78(1985) ; ibid. 69, 100(1986).
17. Kuperin Yu.A., Makarov K.A., Merkuriev S.P., Pavlov B.S., Motovilov A.K. Preprint Budapest Univ. ITP-Budapest-Report No 441, Budapest, 1986.
18. Kuperin Yu.A., Kvitsinsky A.A., Merkuriev S.P., Yarevsky E. Proc. VIII Inter. Semin. on High Energy Problems, Dubna,1987.
19. Kvitsinsky A.A., Kuperin Yu.A., Merkuriev S.P., Motovilov A., Yakovlev S.L. Elem. Part. and Atom. Nucl. 17, 267(1986) (in Russian).
20. Merkuriev S.P., Motovilov A.K. Lett. Math. Phys. 7, 497(1983) ; Preprint Inst. des Sciences Nucléaires de Grenoble ISN 83.41, 1983.
21. Vladimirov V.S. Distributions in Mathematical Physics, Moscow, Nauka, 1979 (in Russian).
22. Merkuriev S.P., Faddeev L.D. Quantum Scattering Theory for Few-Body Systems, Moscow, Nauka, 1985 (in Russian).
23. Birman M.S., Solomjak M.Z. The Spectral Theory of Selfadjoint Operators in the Hilbert Space, Leningrad, Leningrad Univ. Press, 1980 (in Russian).
24. Faddeev L.D. Trudy Matem. In-ta AN SSSR 69,1(1963).
25. Krein M.G. Doklady AN SSSR 52, 657(1946).

ON THE POINT INTERACTION OF THREE PARTICLES

R.A.Minlos
Moscow State University, Moscow, Lenin s Hill

The problem of point interaction of three indistinguishable particles (Bosons) has been studied by several authors K.A. Ter-Martirosjan, G.V.Skornyakov [1] , G.S.Danilov [2], R.A.Minlos and L.D.Faddeev [3] ,[4] .

The Hamiltonian of the three-particle system was treated in [3] ,[4] by means of the theory of self-adjoint extensions of symmetric operators see, for example,[5] . Such an approach to the study of point interactions of two particles has been proposed first by F.A.Berezin and L.D.Faddeev [6]. In this note we consider the general case of point interaction of three distinguishable particles.

The Hamiltonian of such a system should be according to [3] defined as some self-adjoint extension of a symmetric operator

$$H = -(1/2m_1)\Delta_{x_1} \; -(1/2m_2)\Delta_{x_2} \; -(1/2m_3)\Delta_{x_3} \tag{1}$$

acting in the space $L_2 = L^2((R^3)^3)$ of the wave functions $\Psi(x_1,x_2,x_3)$ $x_i \in R^3$ with the domain

$$D_H = \{\Psi \in L_2 \; : \Delta_{x_i}\Psi \in L_2 \; , \Psi|_{\Gamma_{ij}} = 0\} \tag{2}$$

where $\Gamma_{ij} \subset (R^3)^3$ are the hyperplanes

$$\Gamma_{ij} = \{(x_1,x_2,x_3) \in (R^3)^3 \; : x_i = x_j \, , 1 \leq i < j \leq 3\} \tag{3}$$

The domain of the operator H^* consists of the functions $\Psi(x_1,x_2,x_3)$ which

(a) are smooth outside any Γ_{ij}

(b) in a neighbourhood of Γ_{ij} behave asymptotically as

$$\Psi(x_1,x_2,x_3) \approx \frac{A_{ij}(\xi)}{|x_i - x_j|} + B_{ij}(\xi) \quad , x_i,x_j \longrightarrow x \tag{4}$$

where $\xi = (x_1, x_2, x_3)|_{x_j = x_i = x} \in \Gamma_{ij}$ and A_{ij}, B_{ij} are functions on Γ_{ij} in general distributions are allowed; an exact description in terms of Fourier transforms will be given below . It is natural to consider the following symmetric extensions of the operator H. Let $H^*_{\{\varepsilon_{ij}\}} \equiv H^*_\varepsilon$ be the contraction of the operator H^* to the domain

$$D_{H^*_\varepsilon} = \{\Psi \in D_{H^*} : B_{ij} = \varepsilon_{ij} A_{ij},\ 1 \le i < j \le 3\} \qquad (5)$$

where $\varepsilon = \{\varepsilon_{ij}\}$ are real constants. The operators H^*_ε are adjoints to some symmetric extensions H_ε of the operator H. These symmetric extensions of H - we call them Ter-Martirosian-Skornyakov extensions are natural from many points of view. In particular, they are direct analogues of the Hamiltonians of two-particle point interaction [6].

As we shall see below, however, the Ter-Martirosian-Skornyakov extensions are not self-adjoint: the operator H_ε have non-zero deficiency spaces and apparently all their self-adjoint extensions (translation-invariant) are unbounded from below. This phenomenon has been discovered first by the author of [2] and [3] for indistinguishable particles. In the case, however, when only two particles are identical fermions and the third particle is different from them, the operator $H_\varepsilon^{(a),(1,2)} = H_\varepsilon \mid L^a_{2,(1,2)}$ turns out to be self-adjoint and below bounded (here $L^a_{2,(1,2)}$ is the space of wave functions antisymmetric in the variables x_1, x_2) . This result has been established by the present author together with M.Shirmatov [8] . In the present note, we give a detailed formulation of the results together with a concise derivation.

Passing to the Fourier transform of the wave functions

$$\Psi(x_1, x_2, x_3) \longrightarrow \varphi(p_1, p_2, p_3)\ , \quad p_i \in R^3$$

and introducing the total momentum

$$P = p_1 + p_2 + p_3$$

we rewrite the operator H to the form

$$H = (1/2M)\, P^2 + h \qquad (6)$$

It acts on the tensor product $L_2(R^3) \otimes L_2(\hat{\Gamma}, dv)$ where $\hat{\Gamma} \subset (R^3)^3$

is the manifold

$$\hat{\Gamma} = \{(k_1, k_2, k_3) \in (R^3)^3 : k_1 + k_2 + k_3 = 0\}$$

equipped with the natural measure dv.

The quantity M in (6) is defined as $M = m_1+m_2+m_3$. The self-adjoint operator $(1/2M)P^2$ acts in $L_2(R^3)$ while the symmetric operator h acts in $L_2(\hat{\Gamma}, dv)$ as

$$(hf)(k_1,k_2,k_3) = \left((1/2m_1)k_1^2 + (1/2m_2)k_2^2 + (1/2m_3)k_3^2\right) f(k_1,k_2,k_3) \qquad (7)$$

Its domain is of the form

$$D_h = \left\{ f \in L_2(\hat{\Gamma}, dv) : hf \in L_2(\hat{\Gamma}, dv), \int_{\hat{\Gamma}} f(k_1,k_2,k_3)u(k_i)dv = 0, i=1,2,3 \right\} \qquad (8)$$

where the function $u(k)$ fulfils the condition

$$\int_{\mathbb{R}^3} \frac{u(k)^2}{(k^2+1)^{1/2}} d^3k < \infty \qquad (9)$$

The domain D_{h^*} of the adjoint operator h^* consists of functions of the following form (see [3])

$$f(k) + \frac{\sum u_{j,k}(k_i)}{\frac{1}{2m_1}k_1^2 + \frac{1}{2m_2}k_2^2 + \frac{1}{2m_3}k_3^2 + 1} + \qquad (10)$$

$$+ \frac{\sum v_{j,k}(k_i)}{\left(\frac{1}{2m_1}k_1^2 + \frac{1}{2m_2}k_2^2 + \frac{1}{2m_3}k_3^2 + 1\right)^2}$$

where $f \in D_h$ and $u_{j,k}$, $v_{j,k}$ belong to the class (9) and h^* acts according to the formula

$$(h^* g)(k_1,k_2,k_3) = \left(\frac{1}{2m_1}k_1^2 + \frac{1}{2m_2}k_2^2 + \frac{1}{2m_3}k_3^2\right) g(k_1,k_2,k_3) -$$

$$- \sum u_{j,k}(k_i) \qquad . \qquad (11)$$

The element $g \in D_{h^*}$ determined in the representation (10) by a rapidly decreasing function $u_{j,k}$ has the following asymptotic behaviour

$$\int_{\substack{\hat{\Gamma}^s_{i,j}\cap\{|K_i|,|K_j|<N\} \\ 1\le i<j\le 3}} g(k_1,k_2,k_3)\, dv_s^{i,j} = 4\pi\mu_{i,j}u_{i,j}(s)N + B_{i,j}(s) + O(1) \ , \ N\to\infty \qquad (12)$$

where

$$\hat{\Gamma}^s_{i,j} = \left\{ k_i + k_j = -s \right\} \subset \hat{\Gamma}$$

is a 3-dimensional submanifold of $\hat{\Gamma}$ (equipped with the measure dv_s) and $B_{i,j}$ belongs to the class (9). The aforementioned extension H_ε of the operator H is obtained by an extension h_ε of h the adjoint h_ε^* of which is defined on the domain

$$B_{i,j} = 4\pi\mu_{i,j}\varepsilon_{i,j}\, u_{i,j}; \ 1\le i<j\le 3 \qquad (13)$$

We shall now compute the resolvent $(h_\varepsilon - zE)^{-1}$ of the operator h_ε . Setting $f = (h_\varepsilon - zE)^{-1}g$ we find from (11) that

$$f(s) = \frac{g(s) + \sum u_{i,j}(s_k)}{\frac{1}{2m_1}s_1^2 + \frac{1}{2m_2}s_2^2 + \frac{1}{2m_3}s_3^2 - z} \ ; \ s = (s_1,s_2,s_3)\in\hat{\Gamma} \qquad (14)$$

According to (12)-(14) the functions $u_{i,j}$, $1\le i<j\le 3$ solve the following system of equations

$$\left(4\pi\varepsilon_{i,j}\mu_{i,j} + 2\pi^2(2\mu_{i,j})^{3/2}\sqrt{\frac{M\mu_{i,j}}{m_1m_2m_3}p^2 - z}\right)u_{i,j}(p) -$$

$$- \int_{\hat{\Gamma}^p_{i,j}} \frac{u_{i,k}(s_j) + u_{j,k}(s_i)}{\frac{1}{2m_1}s_1^2 + \frac{1}{2m_2}s_2^2 + \frac{1}{2m_3}s_3^2 - z}\, dv_p^{i,j} = \qquad (15)$$

$$= \chi_{i,j}(p,z) \equiv \int_{\hat{\Gamma}^p_{i,j}} \frac{g(s)}{\frac{1}{2m_1}s_1^2 + \frac{1}{2m_2}s_1^2 + \frac{1}{2m_3}s_3^2 - z}\, dv_p^{i,j}$$

Let T be the symmetric operator in $L_2^3 = L_2(R^3)\oplus L_1(R^3)\oplus L_2(R^3)$

$$(T u)_{i,j}(p) = 2\pi^2(2\mu_{i,j})^{3/2}\left(\sqrt{\frac{M\mu_{i,j}}{m_1m_2m_3}p^2 + 1}\right)u_{i,j}(p) -$$

$$- \int_{\hat{\Gamma}^p_{i,j}} \frac{u_{i,k}(s_j) + u_{j,k}(s_i)}{\frac{1}{2m_1}s_1^2 + \frac{1}{2m_2}s_2^2 + \frac{1}{2m_3}s_3^2 + 1}\, dv_p^{i,j}$$

defined on finite functions .

Using the same arguments as in [3], we find that self-adjoint extensions of h_ε are obtained from extensions of the operator T and that the deficiency indices of h_ε and T are equal.

Theorem 1

The operator T on $L_2^{(3)}$ (and consequently, the operator h_ε on $L_2(\hat{\Gamma}, dv)$) has non-zero, finite and mutually equal deficiency indices.

Proof: The operator T commutes with the operators V_g, $g \in O_3$ of the rotation group representations in $L_2^{(3)}$

$$(V_g u)_{i,j}(p) = u_{i,j}(g^{-1}p)\ ,\quad g \in O_3\ ,\quad u \in L_2^{(3)} \tag{17}$$

and is reduced by every space $L_2^{(3)}(l)$ whose elements are triples of the functions

$$u_{j,k}^{l}(p) = \sum_m u_{j,k}^{l,m}(|p|)\, Y_m^l(\Theta,\varphi) \tag{18}$$

Here $|p|, \Theta, \varphi$ are the spherical coordinates of p, $Y_m^l(\Theta,\varphi)$ are the spherical funcions of the weight l (see [7]) and $u_{i,j}^{l,m}$ are functions of $|p|$.

Let T_l be the part of T acting in the space $L_2^{(3)}(l)$. We shall consider first the case $l = 0$ i.e. the spherically symmetric $u_{i,j}(p)$. By the calculation similar to those of [3] and [4], it is possible to check that the domain of $(T_l)^*$, $l=0$ consists of triples $\{u_{i,j}\}$ of the form

$$u_{i,j}(p) = \frac{\sum C_{i,j}^q\, |p|^{is_q}}{p^2 + 1} + \tilde{u}_{i,j}(p) \tag{19}$$

where $\tilde{u}_{i,j}$ are functions from the domain of the closure $\bar{T}_{l=0}$ and $C_{i,j}^q$ solves the following homogeneous system of equations

$$\left(\frac{m_1 m_2 m_3}{M}\right)^{-1/2} s_q\, \mathrm{ch}\!\left(\frac{\pi s_q}{2}\right) C_{i,j}^q - \frac{m_j}{\mu_{k,j}^2}\left(\frac{\mu_{k,j}}{\mu_{i,j}}\right)^{is_q/2} \mathrm{sh}(\varphi_j s_q)\, C_{k,j}^q -$$

$$- \frac{m_i}{\mu_{k,i}^2}\left(\frac{\mu_{k,i}}{\mu_{i,j}}\right)^{is_q/2} \mathrm{sh}(\varphi_i s_q)\, C_{k,i}^q = 0 \tag{20}$$

where s_q are real zeros of the determinant of this system

$$D\,s = \left(\frac{m_1 m_2 m_3}{M}\right)^{-3/2} \Big\{ s^3 ch^3\left(\frac{\pi s}{2}\right) - \frac{s\, ch\left(\frac{\pi s}{2}\right)}{M m_1 m_2 m_3} \times$$

$$\times \Big[(m_1+m_3)^2 (m_2+m_3)^2 \; sh^2(\varphi_3 s) + (m_1+m_2)^2 (m_3+m_2)^2 \; sh^2(\varphi_2 s) +$$

$$+ (m_2+m_1)^2 (m_3+m_1)^2 \; sh^2(\varphi_1 s)\Big] - \frac{2 \prod (m_i+m_j)^2}{(M m_1 m_2 m_3)^{3/2}} \prod_i sh(\varphi_i s) \Big\} \quad .(21)$$

The angles φ_1, φ_2, φ_3 in (20) are defined by the relations

$$\sin \varphi_i = \left(\frac{m_j m_k}{(m_j+m_i)(m_k+m_i)}\right)^{1/2}, \quad 0 < \varphi_i < \frac{\pi}{2} \quad . \; (22)$$

It is easy to check that the functions $s^{-3}D(s)$ is even, negative-valued at $s = 0$ and positive for $|s| \gg 1$. Thus there are at least two real zeros of the function $D(s)$ and the total number of its zeros is finite. It means that the operator $T_{l=0}$ has non-zero, finite and mutually equal deficiency indices. The operators T_l, $l = 1,2,\ldots$ can be treated in a similar way and moreover it is possible to prove that they are essentially self-adjoint for large enough l. Hence the theorem is proved.

Remark:

The self-adjoint extensions of h_ε may be unbounded from below similarly as it was proved for identical particles bosons in [3] and [4] . In order to prove such an assertion one has to know how the coefficients $C^q_{i,j}$ in the representation (19) for the solution of a system of equations similar to (15) behave assymptotically as $z \to -\infty$ (such a system arises when we are looking for the eigenfunctions of the operator h_ε with negative eigenvalues z). The unboundedness from below of the extensions of h_ε is an unpleasant property, since only below-bounded Hamiltonians have a physical meaning. In the particular case, however, when two particles are identical fermions and the third one is of a different kind, the operator h_ε is self-adjoint and bounded from below. More exactly, let $L^{(a)}_{2,(1,2)} \subset L_2(\hat{\Gamma}, dv)$ be the space of functions on $\hat{\Gamma}$ which are antisymmetric in the variables k_1, k_2, i.e., $\Psi(k_1,k_2,k_3) = -\Psi(k_2,k_1,k_3)$. If $m_1 = m_2$ and $\varepsilon_{12} = 0$, $\varepsilon_{13} = \varepsilon_{23} = \varepsilon$ then this space is invariant with respect to the operator h_ε . Let $h^{(a)}_{\varepsilon,(1,2)}$ be the restriction of h_ε to $L^{(a)}_{2,(1,2)}$, then the following theorem is valid:

Theorem 2:
The operator $h^{(a)}_{\varepsilon,(1,2)}$ on $L^{(a)}_{2,(1,2)}$ is self-adjoint, bounded from below and its discrete spectrum is negative and finite.

The operator $\left(h^{(a)}_{\varepsilon,(1,2)}\right)^{*}$ on $L^{(a)}_{2,(1,2)}$ is defined by the formula

$$\left(h^{(a)}_{\varepsilon,(1,2)}\right)^{*} f = \left(\frac{1}{2m_1} k_1^2 + \frac{1}{2m_2} k_2^2 + \frac{1}{2m_3} k_3^2\right) f - u(k_1) + u(k_2) \qquad (23).$$

In order to write down the resolvent $\left(h^{(a)}_{\varepsilon,(1,2)} - zE\right)^{-1}$ of $h^{(a)}_{\varepsilon,(1,2)}$ one has to find a solution to the following integral equation for the function u which is similar to (15)

$$4\pi\varepsilon\mu_{1,3} + 2\pi^2(\mu_{1,3})^{3/2}\sqrt{\frac{2m + m_3}{m(m + m_3)}\, p^2 - z}\;\; u(p) +$$

$$+ \int_{\mathbb{R}^3} \frac{u(s)\, d^3s}{\left(\frac{1}{2m} + \frac{1}{2m_3}\right) p^2 + \left(\frac{1}{2m} + \frac{1}{2m_3}\right) s^2 + \frac{1}{2m_3}\,(p,s) - z} = \qquad (24)$$

$$= \chi(p,z)\,, \quad m = m_1 = m_2 \quad .$$

The corresponding homogeneous equation has not solutions of the form (19), which implies that $h^{(a)}_{\varepsilon,(1,2)}$ is self-adjoint and bounded from below. A detailed study of the quation (24) is needed to prove the finiteness of the discrete spectrum.

The last result suggests the following interestig problem: what kind of assumptions about statictics can ensure existence of a self-adjoint and below-bounded operator of the form (1) for a general system of n particles with a point interaction? In the case n = 4 for example, there are the following possibilities:

(a) three identical fermions interacting with a fourth particle of a different kind
(b) two different couples of identical fermions interacting with each other
(c) a couple of identical fermions interacting with a couple of bosons.

In conclusion, let us remark that the above described scheme worked out for scalar particles can be extended to the case of particles with spin.

References

1. K.A.Ter-Martirosian, G.V.Skornyakov: Sov.Phys.JETP 31 1956 775
2. G.S.Danilov: Sov.Phys. JETP 40 (1961) 498
3. R.A.Minlos, L.D.Faddeev: Dokl.Acad.Sci.USSR 141 (1961) 335
4. R.A.Minlos, L.D.Faddeev: Sov.Phys. JETP 41 (1961) 1850
5. M.Reed, B.Simon: Methods of Modern Mathematical Physics, vol.II Academic Press, New York, 1975
6. F.A.Berezin, L.D.Faddeev: Dokl.Acad.Sci. USSR 137 (1961) 1011
7. I.M.Gelgand, R.A.Minlos, Z.Ya.Shapiro: Representations of the rotation group and Lorentz group. Fizmatgiz, Moscow 1959 (in Russian)
8. R.A.Minlos: in " Quantum Field Theory and Quantum Statistics. Essays in Honor of the 60-th Birthday of E.S.Fradkin, vol.I, Adam Hilger, Bristol, 1987, p. 393

A RESONATING-GROUP MODEL WITH EXTENDED CHANNEL SPACES

Yu.A.Kuperin, K.A.Makarov, Yu.B.Melnikov
Physical Institute, Leningrad State University,
1 Maya 100, Petrodvoretz Leningrad 198904

A new version of the resonating-group model with extended relative-motion space which takes into account the effect of additional two-body resonant channels is formulated using the theory of self-adjoint extensions. It is shown that after projecting on the original relative-motion space we get an effective non-local energy-dependent interaction which describes Pauli repulsion for small intercluster distances. We propose also a generalization of this scheme for three-cluster systems.

1. Introduction

In this lecture, we propose a new method for exclusion of forbidden states in cluster scattering [1-6] based on the abstract spectral deformation scheme of non-relativistic Hamiltonians [7-12]. In distinction to the standard resonating-group model (RGM), the exclusion of Pauli-forbidden states is achieved not by truncation of the relative-motion space, but using an extension of the standard RGM space $\mathcal{H}^{RGM} = A\{\mathcal{H}_1^{in} \otimes \mathcal{H}_2^{in} \otimes \mathcal{H}^0\}$. Here $\mathcal{H}_a^{in}$ are the spaces referring to the internal-cluster degrees of freedom, $\mathcal{H}^0$ is the relative-motion space and A is the antisymmetrization operator. The extended space $A\{\mathcal{H}_1^{in} \otimes \mathcal{H}_2^{in} \otimes [\mathcal{H}^0 \oplus \mathcal{H}^1]\}$ includes the dynamics of compound resonances, which can arise when the intercluster distance is small enough to allow overlapping of clusters. This dynamics is described by an additional Hilbert space $\mathcal{H}^1$. Such extension of the space $\mathcal{H}^{RGM}$ makes it possible to construct a total self-adjoint (s.a.) operator which describes the common dynamics of all degrees of freedom of the system starting from the Hamiltonians acting in the channels $\mathcal{H}^b$, b = 0, 1. The construction is based on the theory of s.a. extensions, and enables to obtain an effective description of the class of effective energy-dependent interactions $V^{eff}(z)$, arising in $\mathcal{H}^0$ after the exclusion of the resonant channel $\mathcal{H}^1$. As we show, the special choice of spectral characteristics of the operator acting in $\mathcal{H}^1$ leads to the exclusion of one or several forbidden states from the spectrum of the RGM Hamiltonian. Thus the energy-dependent potential $V^{eff}(z)$ imitates Pauli principle in $\mathcal{H}^0$. In distinction to the "operators" of the standard scheme [1-6], the new "Hamiltonian" with the additional effective potential $V^{eff}(z)$ obtained by our method acts in the complete (non-truncated) relative-motion space.

The paper is organized as follows. Sec.2 is devoted to construction of the new Hamiltonian which imitates Pauli principle in two-cluster systems. An illustration of the exclusion of one Pauli-forbidden state in such a system is presented in Sec.3. We demonstrate also the connection between the presence of a linear term in the potential $V^{eff}(z)$ as a function of energy and the asymptotical "switching off" of Pauli principle in the limit of high energies $z \to \infty$ in cluster scattering. We propose also a generalization of this scheme for three-cluster systems (Sec.4) and study the spectrum and the resolvent of the corresponding Hamiltonians (Sec.5). Faddeev equations for such systems are discussed in Sec.6. Finally, Sec.7 contains some conclusions; in particular, the possibility of exclusion of three-body forbidden states is discussed.

2. The Hamiltonian for a two-cluster system

Let us consider binar processes of cluster scattering and bound states in such systems. Dynamics of the interior-cluster degrees of freedom is defined by s.a. operators H_a^{in}, $a = 1, 2$, while the dynamics of relative motion of clusters by a s.a. operator H^0 acting in $\mathcal{H}^0$. We suppose that the operators H_a^{in} acting in Hilbert spaces $\mathcal{H}_a^{in}$ have a purely discrete spectrum. A possible occurance of resonant compound states shall be imitated with the help of an additional Hilbert space $\mathcal{H}^1$ in which a s.a. operator H^1 acts. If there is no connection to the external channel $\mathcal{H}^0$ the operator H^1 generates the dynamics of resonant degrees of freedom. Let us consider the s.a.

operator $H^0 \oplus H^1$ in the space $\mathcal{H}^0 \oplus \mathcal{H}^1$, which gives the independent dynamics in spaces $\mathcal{H}^0$, $\mathcal{H}^1$. In order to "swith in" an interaction between the channels $\mathcal{H}^0$ and $\mathcal{H}^1$ in accordance with general scheme [7-12], one must restrict the s.a. operators H^0, H^1 to symmetric ones H_0^0, H_0^1 and then construct all s.a. extensions of the operator $H_0^0 \oplus H_0^1$ in the space $\mathcal{H}^0 \oplus \mathcal{H}^1$. Each s.a. extension can be interpreted as a total Hamiltonian of the whole system including the resonant degrees of freedom.

Realization of this scheme includes the following steps:

1. Let s.a. operators H^b, $b = 0, 1$ are defined on domains $D(H^b) \subset \mathcal{H}^b$ and let linear envelope $\mathcal{N}_b = \mathcal{L}\{\theta_k^b\}_{k=1}^n$ be generating subspaces of the discrete parts of these operators (θ_k^b are the corresponding generating elements). We restrict H^b to the symmetric operators H_0^b:

$$H_0^b = H^b \restriction_{D(H_0^b)}; \quad D(H_0^b) = \{(H^b - i)^{-1}\psi : \psi \in D(H^b) \cap \mathcal{N}_b^{\perp}\}. \tag{1}$$

Choosing the basis $\{w_{b,k}^{\pm}\}_{k=1}^n$:

$$w_{b,k}^{+} = \frac{1}{2}\left(U_b\,\theta_k^b + \theta_k^b\right) \text{ and } w_{b,k}^{-} = \frac{1}{2i}\left(U_b\,\theta_k^b - \theta_k^b\right)$$

with

$$U_b = \left(H^b - i\right)\left(H^b + i\right)^{-1}$$

in linear envelope $\mathcal{L}\{\mathcal{N}_b, \mathcal{N}_b^*\}$ (where $\mathcal{N}_b^* = U_b^*\mathcal{N}_b$) one can describe the domain $D(H_0^{b*})$ of the adjoint operator H_0^{b*} by the formula

$$D(H_0^{b*}) = \Big\{u_b = \tilde{u}_b + \sum_{k=1}^{n}\left(\mathcal{E}_{b,k}^{+}(u_b)\,w_{b,k}^{+} + \mathcal{E}_{b,k}^{-}(u_b)\,w_{b,k}^{-}\right), \quad \tilde{u}_b \in D(H_0^b)\Big\}. \tag{2}$$

The domain $D(H_0^b)$ which appears in (1) is not dense in $\mathcal{H}^b$, and therefore the operator H_0^{b*} does not exist in general. One can overcome this difficulty [7,8] by defining the action of H_0^{b*} on the elements $w_{b,k}^{\pm}$ as

$$H_0^b\, w_{b,k}^{\pm} = \mp\, w_{b,k}^{\mp}. \tag{3}$$

Hence

$$H_0^b = H^b\restriction_{D(H_0^b)} + \left(-iI\restriction_{\mathcal{N}_b} + iI\restriction_{\mathcal{N}_b^*}\right)\restriction_{\mathcal{L}\{\mathcal{N}_b, \mathcal{N}_b^*\}} \tag{4}$$

2. The self-adjoint extensions H of the operator $H_0^0 \oplus H_0^1$ in the space $\mathcal{H}^0 \oplus \mathcal{H}^1$ are specified by fixing the boundary conditions (7) which ensure vanishing of the boundary form $\langle H_0^* \mathcal{U}, \mathcal{V}\rangle - \langle \mathcal{U}, H_0^* \mathcal{V}\rangle$:

$$\begin{pmatrix} \mathcal{E}_1^+(u_1) \\ \mathcal{E}_0^-(u_0) \end{pmatrix} = \begin{pmatrix} B_1 & B \\ B^* & B_2 \end{pmatrix} \begin{pmatrix} \mathcal{E}_1^-(u_1) \\ -\mathcal{E}_0^+(u_0) \end{pmatrix} \tag{5}$$

Here $H_0 = H_0^0 + H_0^1$ with $\mathcal{U} = \{u_b\}_{b=0}^1$, $\mathcal{V} = \{v_b\}_{b=0}^1$, and furthermore $B_j = B_j^*$, $j = 1, 2$, and B are $n \times n$-matrices, which play the role of parameters of the model. Vectors $\mathcal{E}_b^{\pm}$ have the components $\mathcal{E}_{b,k}^{\pm}$.

Using the connection between the vectors $\mathcal{E}_b^{\pm}$,

$$\mathcal{E}_{b,k}^-(u_b) = \langle u_b, \phi_k^b \rangle - \sum_{s=1}^{n} \mathcal{E}_{b,s}^+(u_b) \langle H^b \theta_s^b, \theta_k^b \rangle, \tag{6}$$

$$\phi_k^b = (H^b + i)\, \theta_k^b$$

and the relation (4), one can write the operator H defined by the boundary conditions (5) in the matrix form

$$H = \begin{pmatrix} H_V^0 & Z \\ Z^* & H_V^1 \end{pmatrix} . \tag{7}$$

Here $H_V^b = P_b H P_b$, where P_b are the projections on $\mathcal{H}^b$ in $\mathcal{H} = \mathcal{H}^0 \oplus \mathcal{H}^1$. In accordance with (5)-(7), the blocks of the Hamiltonian H are :

$$\begin{aligned} H_V^0 &= H^0 + V = H^0 - \sum_{k,s=1}^{n} J_{ks}^0 \langle \cdot, \phi_s^0 \rangle \phi_k^0 , \\ H_V^1 &= H^1 - \sum_{k,s=1}^{n} J_{ks}^1 \langle \cdot, \phi_s^1 \rangle \phi_k^1 , \\ Z &= - \sum_{k,s=1}^{n} \tilde{J}_{ks} \langle \cdot, \phi_s^1 \rangle \phi_k^0 , \end{aligned} \tag{8}$$

where J^0, J^1, $\tilde{J}$ are the following matrices

$$J^0 = \left[K_0 + B^* (I - K_1 B_1)^{-1} K_1 B - B_2 \right]^{-1} ,$$

$$J^1 = \left[B^* K_1 - (B_2 - K_0) B^{-1} (B_1 K_1 + I) \right]^{-1} \left[B^* - (B_2 - K_0) B^{-1} B_1 \right] ,$$

$$\tilde{J} = -J^0 B^* (I + K_1 B_1)^{-1} .$$

The matrices K_b, b=0,1, are defined as follows

$$(K_b)_{ij} = \langle H^b \phi^b_j, \phi^b_i \rangle .$$

Let a matrix Δ^b connect the vectors $\mathcal{E}^{\pm}_b$:

$$\mathcal{E}^-_b = \Delta^b \mathcal{E}^+_b \tag{9}$$

Using this relation one can exclude the resonant channel $\mathcal{H}^1$ and obtain from (5)-(9) closed formulas for $P_0 H$:

$$P_0 H = H^0 + V^{eff}(z) = H^0 - \sum_{k,s=1}^{n} (D_0^{-1})_{ks} \langle \cdot , \phi^0_s \rangle \phi^0_k \tag{10}$$

where the matrix D_0 has the following form

$$(D_0)_{ij} = \Delta^0_{ij} + \langle H^0 \theta^0_i, \theta^0_j \rangle . \tag{11}$$

On the eigensubspaces $(H - z)\mathcal{U} = 0$ the matrix elements $\Delta^b_{ij}(z)$ depend on total energy in the c.m. frame and are equal to the Schwartz integrals of the spectral measure $d\langle \mathcal{S}^b_\lambda \theta^b_i, \theta^b_j \rangle$ of the s.a. operator H^b (see [7]) :

$$\Delta^b_{ij}(z) = \langle (I + zH^b)(H^b - z)^{-1} \theta^b_i, \theta^b_j \rangle = \\ = \int_R \frac{1 + \lambda z}{\lambda - z} d \langle \mathcal{S}^b_\lambda \theta^b_i, \theta^b_j \rangle . \tag{12}$$

The external $\mathcal{H}^0$ and resonant $\mathcal{H}^1$ channel are coupled together by the formula

$$\Delta^0 = B^* \Delta^1 (B_1 \Delta^1 - I)^{-1} B - B_2 , \tag{13}$$

which is a simple consequence of (5), (9).

We have found that our scheme of adding the resonant channel $\mathcal{H}^1$ leads automatically to existence of an additional energy-dependent potential $V^{eff}(z) = - \sum (D_0^{-1})_{ks} \langle \cdot , \phi^0_s \rangle \phi^0_k$ in the channel $\mathcal{H}^0$. This separable non-local interaction imitates antisymmetrization effects and, as we show below, leads to exclusion of the Pauli-forbidden states from the channel $\mathcal{H}^0$. The parameters $(D_0^{-1})_{ks}$ (i.e. matrix elements $\Delta^1_{ij}(z)$) are determined by the positions of compound resonances and can be fitted from the scattering data.

The resolvent R(z) of the total Hamiltonian H has the matrix form $\{R_{ab}(z)\}$, where $R_{ab}: \mathcal{H}^b \to \mathcal{H}^a$, and can be reconstructed from its "external" block $G(z) \equiv R_{00}(z)$ alone [10,11]. Since the effective potential $V^{eff}(z)$ is separable, this block can be easily calculated:

$$G(z) = G_0(z) + \sum_{k,s=1}^{n} (D_0^{-1})_{ks} \sum_{q=1}^{n} \left(N^{-1}(z)\right)_{sq} \left\langle G_0(z)\,\bullet\,,\ \varphi_q^0 \right\rangle \varphi_k^0, \quad (14)$$

where $G_0(z) = (H^0 - z)^{-1}$ and the matrix N(z) is defined by its matrix elements

$$N_{ij}(z) = \delta_{ij} - \sum_{k=1}^{n} (D_0^{-1})_{kj} \left\langle G_0(z)\varphi_k^0,\ \varphi_i^0 \right\rangle .$$

In accordance with (9), the discrete spectrum of the operator H defined by the boundary conditions (5) can be obtained as the set of the solutions to the dispersion equation

$$\det \begin{pmatrix} B_1 - (\Delta^1(z))^{-1} & B \\ B & B_2 + \Delta^0(z) \end{pmatrix} = 0 \quad (15)$$

where $\Delta^b(z)$, b=0,1, are to be calculated from (12).

3. Two-cluster system: an illustration

Analyzing the equation (15), we find that choosing parameters of the model in a proper way, one can exclude effectively a finite number of forbidden states. We are going to demonstrate here this phenomenon in the simplest situation with one forbidden state, using $n = 1$, $B = \beta \in \mathbb{C}$ and $B_1 = B_2 = 0$. In this case, the equation (15) turns into

$$\Delta^0(z) = -\alpha^{-1}\Delta^1(z) \quad (16)$$

where $\alpha = |\beta|^2$ is a real parameter playing the role of a coupling constant of the channels $\mathcal{H}^0$ and $\mathcal{H}^1$.

The above described scheme of coupling the operators H^0 and H^1 employs Schwartz integrals of the form (12) as main parameters. They do not contain linear energy-dependent terms. However, the Riesz-Herglotz theorem allows to extend the class of Schwartz integrals to functions of the following type :

$$\Delta^b(z) = \int_{\mathbb{R}} \frac{1+\lambda z}{\lambda - z}\, d\mu_b(\lambda) + c_1^b z + c_2^b\,, \quad c_1^b \geqslant 0,\ c_2^b \in \mathbb{R} \tag{17}$$

Henceforth we shall use this more general representation for $\Delta^1(z)$ assuming $c_1^0 = c_2^0 = 0$ if $b = 0$. The role of the linear energy-dependent terms was discussed in Refs. 9, 13.

Let thus H^1 be a one-dimensional operator so that

$$\Delta^1(z) = (1+zz^1)(z^1-z)^{-1}\langle \mathcal{P}^1\theta^1,\ \theta^1\rangle + c_1^1 z + c_2^1$$

and let χ_0 be a forbidden state in the channel $\mathcal{H}^0$ with the energy z_0, $(H^0 - z_0)\chi_0 = 0$. Let us choose further a resonance level z^1 in channel $\mathcal{H}^1$ which equals to z_0: $z^1 = z_0$. Then (16) turns into

$$\sum_s \frac{1+zz_s}{z_s - z}\langle \mathcal{P}_s^0\theta^0,\ \theta^0\rangle = -\alpha^{-1}\Big\{c_2^1 + c_1^1 z + \\ + \frac{1+z_0 z}{z_0 - z}\langle \mathcal{P}^1\theta^1,\ \theta^1\rangle\Big\} \tag{18}$$

where $\sigma_d(H^0) = \{z_s\}$ is the discrete spectrum of H^0. We pick a generating element θ^0 which differs slightly from χ_0 in the following sense :

$$\langle \mathcal{P}_s^0\theta^0,\ \theta^0\rangle = \delta_s\ ;\quad 1-\delta_0 \ll 1 \text{ and } \delta_s \ll 1 \quad \text{for } s\neq 0. \tag{19}$$

Assuming then that the coupling constant α is a small parameter (the weak-coupling limit), we find that the relative shift between the spectra $\{z_s'\}$ and $\{z_s\}$ of the operators H and H^0, respectively, equals in the first-order of the perturbation theory to

$$z_s' - z_s \simeq \alpha\,\delta_s(1+z_s^2)\,\times \\ \times\Big\{\frac{1+z_0 z_s}{z_0 - z_s}\langle \mathcal{P}^1\theta^1,\ \theta^1\rangle + c_1^1 z_s + c_2^1\Big\}^{-1}. \tag{20}$$

The external components u_{0s} of the corresponding eigenfunctions $\mathcal{U}_s = \{u_{bs}\}_{b=0}^{1}$ are

$$u_{0s} = G_0(z_s')\,\Phi^0 = \sum_q \frac{\delta_q}{z_q - z_s}\,\chi_q \tag{21}$$

where γ_q are the eigenfunctions of the operator H^0, $(H^0 - z_q)\chi_q=0$.

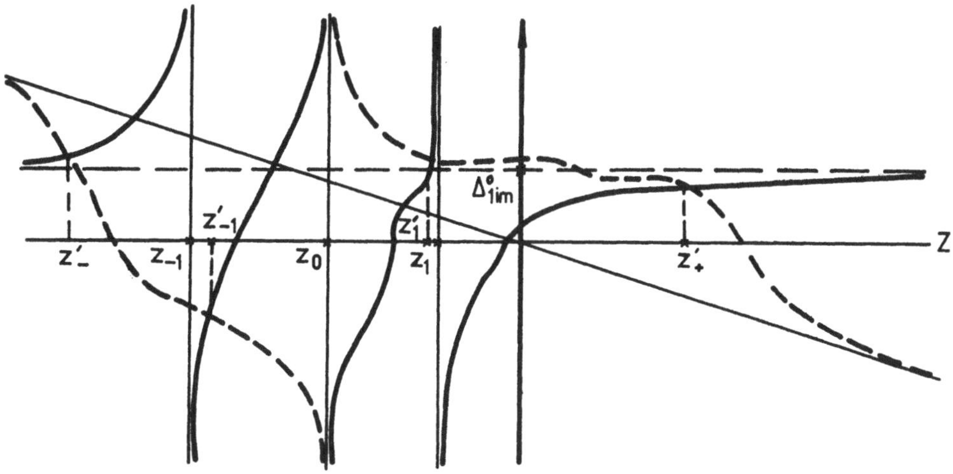

Figure 1: Qualitative solution to the equation (18). The solid curve represents the function $\Delta^0(z)$, the dashed curve does the same for the function $-\alpha^{-1}\Delta^1(z)$ which has the asymptotics $-\alpha^{-1}C_1^1 z$, while $\Delta^0_{lim} \equiv \lim_{z\to\pm\infty} \Delta^0(z) = -\sum_s z_s \delta_s$.

An analysis of the dispersion equation (18), illustrated on Fig.1, leads to the following conclusions :

1. The energy of the forbidden state $z_0 \in \sigma_d(H^0)$ dissapears from the discrete spectrum of H ; all the remaining eigenvalues $z_s \in \sigma_d(H^0)$ are shifted slightly in accordance with (20). Moreover a new eigenvalue $z'_+ > 0$ appears in the continuous spectrum of H such that value $z'_+ \to \infty$ as the coupling constant $\alpha \to 0$.

2. If $\Delta^1(z)$ contains a linear energy-dependent term, $C_1^1 \neq 0$, the spectrum of the total Hamiltonian H is amended furthermore by an additional level $z'_- < 0$ which is called sometimes a phantom [13]. On the other hand, the equations (10), (11) and (17) exhibit the asymptotic "switching off" the antisymmetrization effects in the high-energy limit, i.e. for $C_1^1 \neq 0$ and $z \to \infty$ the effective potential $V^{eff}(z)$ vanishes. These facts concerning the role of the linear energy-dep-

endent term in the function $\Delta^1(z)$ can be summarised in the alternative: either the spectrum of the total Hamiltonian H contains phantoms at large negative energies or the S-matrix (phase shifts) exhibits an anomalous behaviour at high energies related to the fact that $V^{eff}(z) \not\to 0$ as $z \to \infty$.

4. The Hamiltonian for a three-cluster system

One of the most essential questions in the RGM scheme with additional resonance channels is how to include a two-body energy-dependent effective potential into the three-body Hilbert space. The starting point is to study a three-cluster system with a resonance in a fixed pair α only *). One should know the form of the pair potential generated by such an additional resonance channel in this pair. The Hamiltonian $\mathcal{H}_\alpha$ for the three-cluster system with the resonance interaction in the pair α only can be obtained [14] from the pair Hamiltonian

$$H_\alpha = \begin{pmatrix} H_V^0 & Z_\alpha \\ Z_\alpha^* & H_V^\alpha \end{pmatrix} \tag{22}$$

constructed in Sec.2 (see eq. (7)) by the formula

$$\mathcal{H}_\alpha = H_\alpha \otimes I_{T_\alpha} + I_\alpha \otimes T_\alpha \ . \tag{23}$$

Here T_α is a s.a. operator, acting in a Hilbert space $\mathfrak{T}_\alpha$ which describes an effective non-resonant interaction of the third cluster with the subsystem α **), I_{T_α} and I_α are the identity operators in the spaces $\mathfrak{T}_\alpha$ and $\mathcal{H}_\alpha = \mathcal{H}_\alpha^0 \oplus \mathcal{H}_\alpha^1$, respectively, where the spaces $\mathcal{H}_\alpha^b$ (b=0,1; α fixed) coincide with the spaces $\mathcal{H}^b$ of Sec.2. The operator $\mathcal{H}_\alpha$ is e.s.a. in the space $\mathfrak{h}_\alpha = \mathcal{H}_\alpha \otimes \mathfrak{T}_\alpha$ which can be represented as a direct sum $\mathfrak{h}_\alpha = \mathfrak{h}_\alpha^0 \oplus \mathfrak{h}_\alpha^1$, where $\mathfrak{h}_\alpha^b = \mathcal{H}_\alpha^b \otimes \mathfrak{T}_\alpha$.

*) It should not be confused with the coupling constant $\alpha = |\beta|^2$ of the previous section.

**) In the most simple situation T_α is the kinetic-energy operator.

The elements $\mathcal{U} \in \mathcal{G}_\alpha$ have the components $\{u_a\}$, $a=0,\alpha$, $u_0 \in \mathcal{G}^0_\alpha$, $u_\alpha \in \mathcal{G}^1_\alpha$ and the operator $\mathcal{L}_\alpha$ can be written in the matrix form :

$$\mathcal{L}_\alpha = \begin{pmatrix} H^0_V \otimes I_{T_\alpha} + I_{\alpha,0} \otimes T_\alpha & Z_\alpha \\ Z^*_\alpha & H^\alpha_V \otimes I_{T_\alpha} + I_{\alpha,1} \otimes T_\alpha \end{pmatrix} \tag{24}$$

where $I_{\alpha,b}$ are the identity operators in $\mathcal{H}^b_\alpha$. This relation can be rewritten further as follows

$$\mathcal{L}_\alpha = \mathcal{L}_{\alpha,0} + \hat{V}_\alpha , \tag{25}$$

$$\mathcal{L}_{\alpha,0} = \begin{pmatrix} H^0 \otimes I_{T_\alpha} + I_{\alpha,0} \otimes T_\alpha & 0 \\ 0 & 0 \end{pmatrix} \tag{26}$$

$$\hat{V}_\alpha = \begin{pmatrix} V_\alpha \otimes I_{T_\alpha} & Z_\alpha \\ Z^*_\alpha & H^\alpha_V \otimes I_{T_\alpha} + I_{\alpha,1} \otimes T_\alpha \end{pmatrix} \tag{27}$$

where V_α coincides with the operator V of Sec.2 (see the first one of the relations (8)). The operator $\hat{V}_\alpha$ is interpreted as the pair potential, generated by the resonant channel $\mathcal{H}^1_\alpha$ in the three-cluster system.

Our aim is now to construct the total Hamiltonian of the three-cluster system with additional resonant channels in each pair subsystem. To this purpose we make the following simplifying assumption Suppose that there exist unitary identification operators $\pi_{\alpha\beta}$: $\pi_{\alpha\beta} \mathcal{G}^0_\alpha = \mathcal{G}^0_\beta$ *). It allows us to drop in the following the index α from the notation of the quantities associated with the space $\mathcal{G}^0_\alpha$. When the operators $\pi_{\alpha\beta}$ exist the three-cluster Hamiltonian $\mathcal{L}$ can be defined in the natural way :

$$\mathcal{L} = \mathcal{L}_0 + \sum_\alpha \hat{V}_\alpha \; . \tag{28}$$

The operator $\mathcal{L}$ is e.s.a. in the space

$$\mathcal{G} = \mathcal{G}^0 \oplus \sum_\alpha \oplus \mathcal{G}^1_\alpha .$$

---

*) In particular, if $H^0 = -\Delta_{x_\alpha} + \mathcal{V}(x_\alpha, y_\alpha)$ and $T_\alpha = -\Delta_{y_\alpha}$ (where x_α, y_α are the Jacobi coordinates in $\mathbb{R}^6$), then the operators $\pi_{\alpha\beta}$ exist.

In the $\mathcal{G}$ representation , the Hamiltonian $\mathcal{L}$ is given by the operator-valued matrix $\{\mathcal{L}_{ab}\}_{a,b=0}^{3}$ with the elements $\mathcal{L}_{00} = H^0 \otimes I_T + I_{0,0} \otimes T$, $\mathcal{L}_{\alpha\alpha} = H_V^{\alpha} \otimes I_{T_\alpha} + I_{\alpha,1} \otimes T_\alpha$, $\mathcal{L}_{0\beta} = Z_\beta$, $\mathcal{L}_{\beta 0} = Z_\beta^*$, $\alpha, \beta = 1, 2, 3$; remaining elements equal zero. The last mentioned fact is a consequence of the maximal simplicity of the chosen s.a. extension,

$$\begin{pmatrix} \mathcal{E}_1^+ \\ \mathcal{E}_2^+ \\ \mathcal{E}_3^+ \\ \mathcal{E}_0^- \end{pmatrix} = \begin{pmatrix} B_{11} & 0 & 0 & B_{10} \\ 0 & B_{22} & 0 & B_{20} \\ 0 & 0 & B_{33} & B_{30} \\ B_{10} & B_{20} & B_{30} & B_{00} \end{pmatrix} \begin{pmatrix} \mathcal{E}_1^- \\ \mathcal{E}_2^- \\ \mathcal{E}_3^- \\ -\mathcal{E}_0^+ \end{pmatrix} \tag{29}$$

which corresponds to such a structure of the operator-valued matrix $\mathcal{L}$. This choice supposes existence of a coupling between each resonance channel and the exterior channel, while the resonance channels are uncoupled mutually. This excludes existence of three-body forces generated by an interior structure of the clusters.

5. Spectrum of the three-body Hamiltonian

The spectrum of the Hamiltonian $\mathcal{L}$ can be formally obtained from the dispersion equation

$$\mathrm{Det}_{reg}\,\mathcal{B}(z) = 0 \tag{30}$$

in terms of the s.a. extension parameters B_{ab} (29). Here Det_{reg} denotes the regularized operator determinant, $\mathcal{B} = \{\mathcal{B}_{ab}(z)\}_{a,b=0}^{3}$ is 4x4-operator-valued matrix with the components $\mathcal{B}_{\alpha\alpha} = B_{\alpha\alpha} - Q_\alpha^{-1}(z)$, $\mathcal{B}_{\alpha 0} = B_{\alpha 0}$, $\mathcal{B}_{0\alpha} = B_{\alpha 0}^*$, $\alpha = 1, 2, 3$; $\mathcal{B}_{00} = B_{00} + Q_0(z)$ while all the remaining $\mathcal{B}_{ab} = 0$. Furthermore, $Q_a(z)$, $a = 0, 1, 2, 3$, are $n \times n$-operator-valued matrices with the elements

$$\left(Q_a(z)\right)_{ik} = \left\langle \left(I_a \otimes I_{T_a} + H^a \otimes \left(zI_{T_a} - T_a\right)\right)\left(H^a \otimes I_{T_a} + I_a \otimes \left(T_a - zI_{T_a}\right)\right)^{-1} \theta_i^a,\ \theta_k^a \right\rangle \tag{31}$$

where H^a are s.a. operators in $\mathcal{H}_\alpha^1$ when $a = \alpha$ or in $\mathcal{H}^0$ when $a=0$ and θ_i^a are their generating elements.

A standard way to understand meaning of the dispersion equation (30) is based on the study of the structure and analytical properties of the resolvent $R(z) = (\mathcal{H} - z)^{-1}$ of the total three-body Hamiltonian $\mathcal{H}$. The components $R_{ab}(z)$ of the 4x4-operator-valued matrix $R(z)$ satisfy the equations [10,11]

$$(\mathcal{H}_{00} - z)R_{0b}(z) + \sum_{\gamma=1}^{3} Z_\gamma R_{\gamma b} = \delta_{0b} I_0 , \tag{32}$$

$$Z_\alpha^* R_{0b}(z) + (\mathcal{H}_{\alpha\alpha} - z)R_{\alpha b} = \delta_{\alpha b} I_\alpha, \quad \alpha = 1, 2, 3. \tag{33}$$

As it was shown in Refs.10,11, all the components $R_{ab}(z)$ can be reconstructed from the component $R_{00}(z)$ alone, and therefore the study of the total resolvent $R(z)$ can be reduced to the study of the "external" component $R_{00}(z)$ only. The equation (33) for b=0 yields

$$R_{\alpha 0}(z) = -(\mathcal{H}_{\alpha\alpha} - z)^{-1} Z_\alpha^* R_{00}(z) \quad . \tag{34}$$

Using then the equations (34) and (32), we obtain closed equations for $G(z) \equiv R_{00}(z)$:

$$\left(H^0 \otimes I_T + I_{0,0} \otimes T + \sum_\alpha W_\alpha(z) - z\right) G(z) = 0 \tag{35}$$

where the energy-dependent potentials are

$$W_\alpha(z) = -Z_\alpha (\mathcal{H}_{\alpha\alpha} - z)^{-1} Z_\alpha^* \tag{36}$$

Here $I_{0,0}$ and I_0 are the identity operators in $\mathcal{H}^0$ and $\mathcal{G}^0 = \mathcal{H}^0 \otimes \mathcal{T}$ respectively. The operators $W_\alpha(z)$ can be expressed in terms of two-body quantities, namely the resolvents $(H_v - z)^{-1}$ and $(T_\alpha - z)^{-1}$. Due to the special structure of the operator $\mathcal{H}_{\alpha\alpha}$, the equations (36) turn into

$$W_\alpha(z) = \frac{1}{2i\pi} \oint_{\Gamma_\alpha} V_\alpha(\zeta) (T_\alpha - \zeta + z)^{-1} \, d\zeta \tag{37}$$

where the contour Γ_α encircles the spectrum of H_v^α. The relation (37) has been obtained using the identity

$$(\mathcal{H}_{\alpha\alpha} - z)^{-1} = \frac{1}{2i\pi} \oint_{\Gamma_\alpha} (H_v^\alpha - \zeta)^{-1} (T_\alpha - \zeta + z)^{-1} \, d\zeta \tag{38}$$

6. Faddeev equations

The equation (36) can be rewritten in the following integral form

$$G(z) = \hat{R}(z) - \hat{R}(z) \sum_{\alpha} W_{\alpha}(z) G(z) , \tag{39}$$

where $\hat{R}(z) = (\mathcal{X}_{00} - z)^{-1}$. This equation has a structure of Lippmann-Schwinger equation and is not of the Fredholm type. Hence it should be rearranged. Let us define the Faddeev components $G^{\alpha}(z) = \delta_{\alpha 1}\hat{R} - \hat{R}W_{\alpha}(z)G(z)$, $\alpha = 1, 2, 3$. According to the relation (39), $G = \sum_{\alpha} G^{\alpha}$ holds which allows to obtain Faddeev equations for the operators $G^{\alpha}(z)$ using the standard procedure [14]:

$$G^{\alpha}(z) = \delta_{\alpha 1} G_{\mathcal{X}_{\alpha}}(z) - G_{\mathcal{X}_{\alpha}}(z) W_{\alpha}(z) \sum_{\gamma \neq \alpha} G^{\gamma}(z) \tag{40}$$

where $G_{\mathcal{X}_{\alpha}}(z) = (\mathcal{X}_{\alpha} - z)^{-1}$ is the resolvent of the operator (25).

One must check that the equation (40) is a Fredholm one. The proof does not differ from the standard one [14] and can be found in Refs.10,11.

7. Conclusions

The system of the equations (40)can serve as a mathematically correct base for studying the bound states and scattering in the three-cluster system with exclusion of two-body Pauli-forbidden states in the two-body subsystems. This exclusion is connected with the possible occurence of two-body compound resonances and differs from the schemes proposed earlier which restricted the relative-motion space [2-5].

The relations (32), (33) provide us with a solid justification for the equation (40) from the operator point of view which can be hardly obtained in the models with restricted relative-motion space.

One should remark also, that (40) is valid in a limited region of energies only, i.e. below the break-up threshold of individual cluster, because it does not take into account the reaction channels.

The equations (40) can not be used also for exclusion of three-body forbidden states, which can exist in the system. However, these states can be easily excluded by adding an additional orthogonal channel $\mathcal{G}_*$ to the space $\mathcal{G}$,which imitates the possible occurence of three-body resonances. The operators $\mathcal{L}$, $\mathcal{L}_*$ acting in the spaces $\mathcal{G}$ and $\mathcal{G}_*$ can be coupled also by the extension-theory methods in a way similar to the one described above.

Acknowledgements

We are grateful to B.S.Pavlov and S.P.Merkuriev for useful discussions.

References

1. K.Wildermuth, Y.C.Tang: A Unified Theory of the Nucleas, Vieweg, Braunschweig 1977.
2. S.Saito, S.Okai, R.Tamagaki, M.Yasuno, Prog. Theor. Phys. 50 (1973), 1561.
3. S.Saito, Prog. Theor. Phys. Suppl. 62 (1977), 11.
4. E.W.Schmid, Z. Phys. A297 (1980), 105.
5. V.I.Kukulin, V.G.Neudatchin, V.V.Pomerantsev, Yad. Fiz. 24 (1976), 298.
6. B.Buck, H.Friedrich, C.Wheatly, Nucl. Phys. A 275 (1977), 246.
7. B.S.Pavlov, Teor. Mat. Fiz. 59 (1984), 357.
8. Yu.A.Kuperin, K.A.Makarov, B.S.Pavlov, Teor. Mat. Fiz. 63 (1985), 78.
9. Yu.A.Kuperin, K.A.Makarov, B.S.Pavlov, Teor. Mat. Fiz. 69 (1986), 100.
10. Yu.A.Kuperin, K.A.Makarov, S.P.Merkuriev, A.K.Motovilov: in "Theory of Quantum Systems with Strong Interactions", Kalinin 1987; p.4.
11. Yu.A.Kuperin , K.A.Makarov, S.P.Merkuriev et al., ITP-Budapest, Report N 441, Budapest 1986; Proc. of the 3rd School on Few-Body and Quark-Hadron Interactions, Vilnius 1986, Part 2; p.28.
12. B.S.Pavlov, Usp. Mat. Nauk 42 (1987),6(258), 99.
13. I.L.Gratch, Yu.S.Kalashnikova, I.M.Narodetskii et. al., Yad. Fiz. 42 (1985), .
14. S.P.Merkuriev , L.D.Faddeev: Quantum Scattering Theory in Few-Body Systems, Nauka, Moscow 1985 (in Russian).

THE PROBLEM OF A FEW QUASI-PARTICLES IN SOLID-STATE PHYSICS

Alexandr I. Mogilner
Institute of Metal Physics, Academy of Sciences,
Ural Branch, Sverdlovsk, Kovalevskaya street, 18, USSR

Abstract

The problem of energy spectrum of a few quasi-particles in a crystal is investigated. The form of the N-magnon Hamiltonian in a Heisenberg ferromagnet is obtained and general spectral properties of the Hamiltonian as a cluster operator are demonstrated. The quasiparticle spectrum in the strong coupling limit, the Efimov effect, the current and noncurrent bound states are also discussed.

1. The Hamiltonian of a N-magnon System in a Heisenberg Ferromagnet

Some problems in solid-state physics require a investigation of a few quasi-particles in a crystal. One of the most important problems of this type is the spectral analysis of the Hamiltonian describing a few quasi-particles on a lattice [1]. An important example is, for instance, the N-magnon system in a Heisenberg model.

Let us consider a d-dimensional cubic lattice Z^d each site of which is occupied by a spin. One of the most simple Hamiltonians of this system has the form:

$$H = -\frac{1}{2} \sum_{m,n} J_{m,n} \left(S_m^z S_n^z + b S_n^- S_n^+ \right) + D \sum_n \left(S_n^z \right)^2 \qquad (1.1)$$

where $J_{m,n}$ are the exchange integrals, $J_{m,n} = J_{m-n} = J_r$, $J_r \geqslant 0$, $J_r < C_1 \exp(-C_2 |r|)$, $C_1, C_2 > 0$, S_n^z is the operator of spin projection on the z axis at the n-th site, S_n^- (S_n^+) is the operator lowering (rising) the z component of the spin by one (see [2]), b and D are exchange and single-ion anisotropy coefficients, $0 \leqslant b \leqslant 1$, $D \leqslant 0$.

Since the operator $S_0 = \sum_n S_n^z$ of the total spin projection on the z axis commutes with the Hamiltonian H, it is possible to classify the stationary states using its eigevalues. The highest eigenvalue of S_0 corresponds to the ground state while the excited states correspond to the eigenvalues of S_o, which are by 1,2,...,N less than the highest one. The wavefunctions corresponding to these states are given by

$$x_{\{n\}} = A_g^{-1} \prod_{i=1}^{N} S_{n_i}^{-} x_o \qquad (1.2)$$

where A_g is the normalisation coefficient, g is the multiindex ascribed to the set of sites $\{n\} = \{n_1 \ldots n_N\}$, $g = \{g_1 \ldots g_s\}$ and g_m is the number of sites with m spin flips,

$$s \leqslant N, \quad \sum_{m=1}^{s} m g_m = N$$

$$A_g = \prod_{m=1}^{s} (A_m)^{g_m/2} \quad , \quad A_m = \prod_{l=o}^{m-1} (2S-l)(l+1) \qquad (1.3)$$

It is clear that if $g_m \neq 0$ for $m > 2S$, then $x_{\{n\}}=0$, i.e., there can not be more than 2S spin flips at one site. Moreover the spin operators corresponding to different sites commute and this implies that the functions $x_{\{n\}}$ are symmetric with respect to variable permutations.

The functions (1.2) are not eigenfunctions of the Hamiltonian (1.1). The wavefunctions corresponding to the N-magnon stationary states can be constructed as a linear combination of the functions $x_{\{n\}}$:

$$x = \sum_{\{n\}} Q_g^{-1} f_{\{n\}} x_{\{n\}} \qquad (1.4)$$

The constants Q_g are choosen in such a way that

$$|x|^2 = \sum_{\{n\}} |f_{\{n\}}|^2$$

where we are summing over all different functions $f_{\{n\}}$.

$$Q_g = \left(N! \prod_{m=1}^{s} (m!)^{-g_m}\right)^{1/2} \qquad (1.5)$$

Substituting the functions (1.4) into the eigenfunction equation for the Hamiltonian (1.1) and identifying the coefficients of $x_{\{n\}}$ we get after a cumbersome calculations the expression for the Hamiltonian acting in the space of N-magnon functions $f_{\{n\}} \in l_2^{sym}(Z^{dN})$:

$$Hf = FN\, f + bSBf + Vf \qquad (1.6)$$

$$F = D(1-2S) + SJ \; , \quad J = \sum_r J_r \; ,$$

V is the operator of the dynamical two-magnon interaction.

$$(Vf)_{\{n\}} = \Big(\sum_{i\neq j}\big(D\,\delta_{n_i,n_j} - \tfrac{1}{2}\,J_{n_i,n_j}\big)\Big)\, f_{\{n\}} \tag{1.7}$$

B is the operator of kinetic energy and kinematic magnon interaction,

$$(Bf)_{\{n\}} = \sum_{i=1}^{N} \sum_{y} B_n^{i,y}\, f_{\{(n\setminus n_i)\,y\}} \tag{1.8}$$

and $\{(n\setminus n_i)\,y\}$ denotes the set $\{n_1 \ldots n_{i-1}\; y\; n_{i+1} \ldots n_N\}$. The transfer amplitude of the i-th spin flip from the site n_i to the site y is given by

$$B_n^{i,y} = J_{n_i,y}\,\frac{\sqrt{(2S+1-p)\,(2S+1-v)}}{2S} \tag{1.9}$$

where p is the number of spin flips on the site n_i in the configuration $\{n\}$ and v is the number of spin flips on the site y in the configuration $\{(n\setminus n_i)\,y\}$.

Let us note that the expressions (1.6)-(1.9) are correct also for $N>2S$. $B_n^{i,y}=0$ for $p,v>2S$, i.e. the functions $f_{\{n\}}$ with $g_m\neq 0$ for $m>2S$ do not contribute to the r.h.s. of (1.6). The equations for $f_{\{n\}}$ with $\{n\}$ such that $g_m\neq 0$ for $m>2S$ must be considered as absent.

Let us turn from the coordinate to momentum representation:

$$f_{\{n\}} = (2\pi)^{-\frac{dN}{2}} \int_{(T^d)^N} \psi(k_1 \ldots k_N)\,\exp\Big(-i\sum_{j=1}^{N}(k_j, r_{n_j})\Big)\, dk_1 \ldots dk_N \tag{1.10}$$

where $\psi(k_1 \ldots k_N) = \psi(\{k\})$ is the N-magnon wavefunction, $\psi(\{k\}) \in L_2^{sym}((T^d)^N)$, k_j is the quasi-momentum of the j-th magnon, $k_j \in T^d$ (T^d is a d-dimensional torus) and r_{n_j} is the coordinate of the site j. The quasi-momentum k_j is measured in units of a^{-1}, where a is the step of the lattice.

The expressions for the operators B and V have in the momentum representation the following form:

$$(V\psi)(\{k\}) = \sum_{i\neq j} \int_{(T^d)^N} W_2(k_i,k_j,k_i'k_j')\,\delta(k_i+k_j-k_i'-k_j') \times \prod_{l\neq i,j} \delta(k_l-k_l')\,\psi(\{k'\})\, dk_1' \ldots dk_N' \tag{1.11}$$

$$W_2(k_i,k_j,k_i',k_j') = 2D - \tfrac{1}{2}\Big(J\big(\tfrac{1}{2}(k_i-k_j-k_i'-k_j')\big) + J\big(\tfrac{1}{2}(k_i-k_j+k_i'-k_j')\big)\Big) \tag{1.12}$$

$$(B\psi)(\{k\}) = \Big(\sum_{i=1}^{N} e(k_i)\Big)\psi(\{k\}) + \sum_{n=2}^{N}\sum_{t(n)}\int_{(T^d)^N} V_n(k_{t(n)}, k'_{t\,n})\times$$

$$\times\delta(K_{t(n)}-K'_{t(n)})\prod_{j\notin t(n)}\delta(k_j-k'_j)\,\psi(\{k'\})\,dk'_1\ldots dk'_N \qquad (1.13)$$

$$J(k) = \sum_r J_r \exp(i(k,r)) \qquad e(k_i) = J(k_i)$$

is the kinetic energy of the i-th magnon and $t(n)$ and $k_{t(n)}$ are the n-element subsets of the sets $\{1\ldots N\}$ and $\{k_1\ldots k_N\}$ respectively.

$$K_{t(n)} = \sum_{j\in t(n)} k_j \quad , \quad K'_{t(n)} = \sum_{j\in t(n)} k'_j \quad ,$$

$$V_2(k_{t(2)},k'_{t(2)}) = \left(\sqrt{1-\frac{2S-1}{2S}}\right)\left(J(k_i)+J(k_j)+J(k'_i)+J(k'_j)\right) \qquad (1.14)$$

$$V_n(k_{t(n)},k_{t(n)}) = \frac{1}{2}\sum_{p=0}^{n-1}\sum_{t(p,i),\,i\in t(n)} C_{(p-1),(n-p)}\times$$

$$\Big[J\Big(\sum_{j\in t(p,i)}(k_i-k'_j+k_i)\Big) + J\Big(\sum_{j\in t(p,i)}(k'_i-k_j+k'_i)\Big) \quad , n=3,\ldots \qquad (1.15)$$

Coefficients $C_{p,v}$ are given by recursive relations:

$$C_{p,v} = \frac{\sqrt{(2S+1-p)(2S+1-v)}}{2S} - \sum_{p'<p,\,v'<v} C_{p',v'} \quad , \quad C_{11}=1 \qquad (1.16)$$

and the functions $V_n(k_{t(n)},k'_{t(n)})$ are the kernels of the N-magnon kinematic interaction operator.

2. The General Spectral Properties of the N-quasi-particle Hamiltonian on a Lattice as a Cluster Operator

The formulas (1.6), (1.11) - (1.16) define the N-magnon Hamiltonian in the momentum representation. This operator turns out to be a particular case of the additive N-particle cluster operator [3]. There are also other examples of systems of a few quasi-particles on a lattice (magnons, electrons, excitons) the Hamiltonians of which can be written in the form (1.6), (1.11) - (1.16) . (The expressions for the W and V kernels are different for different problems). Owing to this common feature, the following multichannel picture is characteristic for spectra of all Hamiltonians of that type [3].

For any N we can fix the total N-magnon quasi-momentum $K=\sum_{i=1}^{N} k_i$ and consider restriction of the operator H to the subspace of wave-functions corresponding to a fixed K. We denote a restricted operator as $H(K)$ (see [4] for a more detailed analysis of the decomposition of the operator H into the "layers" $H(K)$).

There always exist a main N-particle branche of the continuous spectrum of the operator $H(K)$, which corresponds to the free motion of N noninteracting quasi-particles with quasi-momenta $k_1,\dots,k_N$ and energies

$$\sum_{i=1}^{N} e(k_i) \quad , \quad \sum_{i=1}^{N} k_i = K$$

Moreover there exist a finite or countable number of p-particle branches of the continuous spectrum, $p=2,\dots, N-1$, which correspond to the free motion of p clusters with quasi-momenta $K_1,\dots,K_p$ and energies

$$\sum_{i=1}^{p} E_i^{s_i}(K_i)$$

Here K_i is the quasi-momentum of the centre of mass of the i-th cluster corresponding to the s-th bound state of n quasi-particles, $n=2,\dots,(N+1-p)$. In the case n=1 we have one free quasi-particle. The cluster energy $E_i^{s_i}(K_i)$ is a point of discrete spectrum of the Hamiltonian $H(K_i)$ of the n-quasi-particle system. And finally the operator $H(K)$ can have the discrete spectrum corresponding to N-quasi-particle bound states.

The algebraic structure of the Hamiltonian (1.6), (1.11) - (1.16) is analogous to the structure of Shrödinger operator (SO) acting in the space of N particles in a d-dimensional Euclidean space R^d. There is an important difference however: in the case of SO we have always

$$e(k) = \frac{1}{2m} k^2 \tag{2.1}$$

where m is the particle mass and $k \in R^d$. With this analogy we call the few-quasi-particle Hamiltonian as a discrete Schrödinger operator (DSO).

There are many mathematical difficulties in investigation of the SO which are connected with the noncompactness of the space, unboundedness of SO. On the other hand, the structure of spectral branches is much more simple in the SO case (see[5]) because $e(k)$ is then given by the formula (2.1) while for DSO it can be of an arbitrary form.

The structure of the discrete spectrum σ_{disc} of DSO is also more complicated than in the SO case (σ_{disc} denotes all isolated eigenvalues of finite multiplicity. For bound states embedded in the continuous spectrum see [4],[6]). There is a theorem describing the essential spectrum of the cluster operator which is the analogy of the HVZ theorem for SO [7].

1. In the DSO case the N-particle bound state with energy E(K) need not necessarily exist for all values of centre-of-mass quasi-momentum.

2. The bound state energy E(K) may be higher than the energy corresponding to the continuous spectrum.

3. It is possible that $e(0) < e(k)$, $k \neq 0$ and $E(0) > E(K)$, $K \neq 0$ (see §5 below).

4. In the SO case two weakly interacting particles cannot create a bound state for d=3. This is, however, possible, in the lattice case if $e(k) \sim k^4$ for small k [2].

5. If d=3 in the SO case if and if any two particles out of the three a bound state then the bound state of three particles also exists. There are examples, however, that this does not hold on a lattice. For instance two magnons with infinitesimal nearest neighbours interaction form a bound state while three magnons do not.

There are many defailed studies of the two particle case, but only a few papers investigating the $N \geqslant 3$ case (see [1] and references therein) because of the difficulties connected with analysis of the few-particle Hamiltonians. In the DSO case there are two limit situations which can be investigated: the weakly and strongly interacting quasi-particles. The weak interaction case was discussed in our paper [8].

3. Magnon Bound States in the Strong Coupling Limit

In the strong coupling limit (SCL) which has no analogy for SO the quasi-particles on a lattice are motionless. An example of this situation are the spin flips in Ising model with the nearest-neighbour interaction described by the Hamiltonian (1.1) with

$$S=\frac{1}{2},\quad D=b=0,\quad J_r=J \text{ for } |r|=a,\quad J_r=0 \text{ for } |r|>a$$

Here bSB=0 and H_0= const + V where V is the operator of multiplication by a function in the coordinate representation. It is clear that the wavefunctions $x_{\{n\}}$ appearing in (1.2) and describing the spin flips at sites $n_1 \dots n_N$ in the regime when at most one flip occurs at each site are the stationary states of the Hamiltonian corresponding to the energies

$$E_{\{n\}} = \frac{1}{2} NJz - n_{\{n\}}J \tag{3.1}$$

where $n_{\{n\}}$ is the number of the nearest neighbours pairs in the configuration $\{n\}$ and z is the number of the nearest neighbours to the site (z=2d).

In view of the translational symmetry of the problem it is convenient to introduce another system of basis vectors:

$$x_{\{n'\}}(K) = \sum_R e^{i(K,R)}\, x_{\{n\}} \qquad R=\frac{1}{N}\sum_{i=1}^{N} r_{n_i} \tag{3.2}$$

and the configuration $\{n'\}$is defined by the set of (N-1) vectors

$$\{r_{n_1}-r_{n_2},\dots,r_{n_{N-1}}-r_{n_N}\}$$

$$H_0 x_{\{n'\}}(K) = E_{\{n'\}}x_{\{n'\}}(K) \qquad E_{\{n'\}} = \frac{1}{2} JNz - n_{\{n\}}J$$

We fix the conserved centre-of-mass quasi-momentum K. We will use the term connected for all those configurations where for each spin flip there exist a corresponding spin flip in the nearest neighbour site. All others configurations will be called disconnected.

All states corresponding to the disconnected configurations are infinitely degenerate and the part of the states corresponding to the connected configurations has the same energy. Energies corresponding to these states belong to the essential spectrum δ_{ess}. The remaining part of states which correspond to connected configurations are finite degenerate and their energies lie in δ_{disc}.

Let us now suppose that $0<b\ll 1$. It is easy to show that the ope-

rators $H(b) = H_o + \frac{b}{2} B$ form a Kato's analytic family [4] and that

$$\|B\| < CJzN \quad , \qquad C \sim 1 \tag{3.3}$$

We know from the perturbation theory [4] that under the perturbation with $b \ll 1$ the points of δ_{ess} transform into the segments of δ_{ess} which are possibly limited by a finite or countable number of δ_{disc} points. The quantity δ majoring the dimension of these δ_{ess}-segments and limiting points is equal to

$$\frac{b}{2} \|B\| = \frac{b}{2} CJzN$$

Thus the criterion of the applicability of the perturbation theory is

$$\frac{b}{2} \|B\| \ll \min_{\{n\}_1, \{n\}_2} | E_{\{n\}_1} - E_{\{n\}_2} | \qquad b \ll \frac{2}{zN} \tag{3.4}$$

The perturbation removes partially the degeneration of the points of δ_{disc}. It can be shown that the energies of the N-magnon bound states in SCL are contained in segments $[E_{\{n\}} - C \frac{b}{2} zJ \; ; \; E_{\{n\}} + C \frac{b}{2} zJ]$. Let us note that for large N most energetically advantageous forms are the "cubic drops" with n given by:

$$n = \frac{z}{2} N - \frac{z}{2} N^{\frac{d-1}{d}} - C \, , \qquad C \sim 1$$

$$E_{\{n\}} = C \frac{z}{2} JN^{\frac{d-1}{d}} \tag{3.5}$$

It can be also shown that for $N \gg 1$, $|K| \ll 1$ the following relation holds:

$$E_{\{n\}}(K) = \text{const} + \sum_{i=1}^{d} \frac{K_i^2}{2m_i} \quad , \qquad m_i \sim \left(J\left(bN^{d/(d-1)}\right) N\right)^{-1} \tag{3.6}$$

where K_i are the coordinates of the vector K on the main axes. One can see from the formula (3.6) that the magnon-drop effective mass is essentially nonadditive. This is one more important difference between DSO and SO.

Let us consider now a different model:

$$b=1 \, , \quad S \gg 1 \, , \quad N \ll S \, , \quad SJ \ll |D| \tag{3.7}$$

In this model spin flips are bosons interacting by pair contact potentials on a lattice. Two quasi-particles attract each other with a strength D only if they are at the same site and do not interact when they are localized at different sites. Under the condition (3.7) the Hamiltonian H can be written as

$$H = V' + T' + o(S,D) \tag{3.8}$$

$$\begin{cases} (V'f)_{\{n\}} = \left(\sum_{i \neq j} D\,\delta_{i,j} \right) f_{\{n\}} \\ (T'f)_{\{n\}} = S \sum_{i=1}^{N} \sum_{r} J_r\, f_{\{(n \setminus n_i)\,(n_i + r)\}} \end{cases} \tag{3.9}$$

The wavefunctions $x_{\{n\}}$ from (1.2) are the stationary states of the operator V' with the energies

$$E_{\{n\}} = ND(1-2S) + D \sum_{m=1}^{s} g_m \frac{m(m-1)}{2} \tag{3.10}$$

where g_m were defined above. Under the conditions (3.7) the kinetic energy operator T' can be considered as a small perturbation of the potential energy operator V. All the energy levels (3.10) are infinitely degenerate and transform under the perturbation into segments of δ_{ess} except for the nondegenerate lowest level with the energy

$$E_0 = ND(1-2S) + D\,\frac{N(N-1)}{2}$$

This nondegenerate level corresponds to the bound state of N spin flips at one site.

The N-magnon bound state energy becomes under the perturbation

$$E_0' = ND(1-2S) + D\,\frac{N(N-1)}{2} - z\,\frac{(SJ)^2}{D} + o\left(\frac{(SJ)^2}{D}\right) \tag{3.11}$$

and the effective mass tensor components of this state are given by

$$m_i \sim z\, D^{(N-1)} (SJ)^{-N} \tag{3.12}$$

4. The Efimov Effect for Three Quasi-Particles in a Three-Dimensional Crystal

The existence of the phenomenon discovered by Efimov [9-11] is rigorously proved for a system of three particles interacting by attractive short-range two-body potentials. If there are no two-particle bound states and two of three pairs create a virtual state, then there exists an infinite number of three-particle bound states with energies approaching zero. The asymptotic expression for these energies is given by

$$E_n \sim w_1 \cdot e^{-w_2 n}, \quad n=1,2,\ldots \tag{4.1}$$

where w_1 and w_2 are positive constants depending on masses of the particles and on the two-particle scattering length. It was shown in [9-11] that the Efimov effect is caused fully by the two-particle scattering amplitude singularity at $E \to 0$ and $k_i \to 0$. Under these conditions, the singular part of the homogeneous Faddeev equation $\psi = A(E)\psi$ separates. (Here ψ is the wavefunction of the three-particle system and $A(E)$ is an integral operator in the momentum representation. The eigenfunctions of this equation are the stationary states of the system and E are the corresponding energies. For more detail see [12].) This singular part is non-Fredholm for $E \to 0$. Its kernel $A_s(q,q')$ coincides with the kernel of A at $|q|, |q'| < \Lambda$ and $A_s(q,q')=0$ for $|q|, |q'| > \Lambda$ and Λ small. The characteristic numbers of the operator A_s have an accumulation point at zero and the additional Fredholm term $(A-A_s)$ does not change the accumulation point in agreement with the Weyl's theorem [10].

Let us note that for small $|k_i| < \Lambda$ the quasi-particle kinetic energy is given in the general case by $e(k) = \frac{k^2}{2m} + O(k^4)$ (we consider only the highest-symmetry case). Therefore for $|k_i| < \Lambda$ the three-quasi-particle Hamiltonian (1.6),(1.11)-(1.13) coincides with the Hamiltonian of a three-particle system in an Euclidean spase written in the momentum representation with the exception of the three-particle interaction. The short range three-particle interaction adds, however, only a Fredholm term to the Faddeev equation. Therefore the Faddeev operator corresponding to DSO can be decomposed into the singular part A_s^H which coincides with A_s and the Fredholm term: $A^H = A_s^H + \tilde{A}_s^H$. Thus for the three-quasi-particle system in a three-dimensional crystal the Efimov effect takes place at K=0.

1. In analogy to SO [13] in the case of dimensions d=1,2 or for more than three quasi-particles ($N \geq 4$), the Efimov effect is absent.

2. In the case when two of the three particles are impurities, i.e., $e_1(k)=e_2(k)=\text{const}$ (as an example consider a ferromagnetic crystal with two ferromagnetic or antiferromagnetic impurities) the Efimov

effect exists, i.e., there is infinite number of bound states of a magnon with two impurities. The physical nature of these states is clear: a virtual state of the magnon on one impurity creates a long-range interaction with the second impurity. This second impurity can be localized at any site generating in such a way an infinite number of bound states.

3. In the case when one of the quasi-particles is an impurity (two magnons in the ferromagnetic crystal with one antiferromagnetic impurity) the Efimov effect occur if and only if both the magnon with impurity and two magnons create virtual states. In fact there exists only a finite number of bound states in the limit of noninteracting magnons.

4. The effect is not present if a bound state rather than a virtual state appears at the continuous spectrum threshold and the two-partical scattering length remains finite.

5. The effect is apparently absent if all three quasi-particles are fermions with spin $\frac{1}{2}$ (three electrons in the Hubbard model).

6. Let us mention one more system, where the Efimov effect takes place: three magnons in an anisotropic ferromagnet with such an anisotropic constant that two magnons at K=0 create a virtual state.

The Efimov effect must lead to some observable peculiarities in thermodynamics and in kinetic phenomena . The existence of the Efimov effect for three bosons on a lattice interacting by two-particle contact potentials was announced in [1].

5. Current and Noncurrent Ground States of a Few Quasi-Particles on a Lattice without External Field

It is well known that the energy of a system of a few particles the dynamics of which is defined by SO is given by

$$E = E_o + \frac{K^2}{2M}$$

where M is the total mass and E_o is the internal energy of the system, which does not depend on K. The ground state corresponds to K=0, i.e., the ground state is noncurrent. This is, however, not in the case of quasi-particles on a lattice, where the internal energy depends on the centre of mass quasi-momentum K. This dependence is arbitrary in general and the question is whether the ground state of the system on a lattice without external field is current or noncurrent, i.e., whether it has a quasi-momentum $K \neq 0$ or $K=0$? This question is interesting not only from academical or methodological reasons, but also from the viewpoint of possible applications. It is, moreover, related also to the discussion on macroscopic current states in conductors.

In what follows, we give an example of a one-dimensional system with a current ground state. We formulate conditions under which the ground state of a Hamiltonian describing a system with an arbitrary number of quasi-particles is noncurrent.

Let us note, that if we have only one quasi-particle, then its energy is given by $e(k) = \sum_r J_r \exp(i(k,r))$ and the minimum $\min_k e(k)$ is achieved at $K=0$ if and only if $J_r \leqslant 0$ (J_r is the hopping matrix).

The system is described in the coordinate representation by a wavefunction $\psi_{\{n\}}$. The Hamiltonian can be represented in a form:

$$(H\psi)_{\{n\}} = \sum_{i=1}^{N} \sum_r J_{\{n\}}^{\{(n\setminus n_i)\,(n_i+r)\}} \psi_{\{(n\setminus n_i)\,(n_i+r)\}} + \sum_{i \neq j} V_{i,j} \psi_{\{n\}} \quad (5.1)$$

$$V_{\{(n_i+r)\,(n_j+r)\}} = V_{n_i,n_j} \qquad J_{\{n+c\}}^{\{(n+c\setminus n_i+c)\,(n_i+r+c)\}} = J_{\{n\}}^{\{(n\setminus n_i)\,(n_i+r)\}}$$

The centre-of-mass quasi-momentum is conserved and the wavefunction must have the form

$$\psi_{\{n\}} = e^{i(K,R)} \sum_{\{n'\}} c_{\{n'\}} x_{\{n'\}} \qquad \sum_{\{n'\}} \overline{c_{\{n'\}}} c_{\{n'\}} = 1 \quad (5.2)$$

From the minimax arguments we know, that the ground state energy equals to

$$E_{min} = \min \frac{\langle\psi|H|\psi\rangle}{\langle\psi|\psi\rangle} \quad (5.3)$$

and the ground state wavefunction is described by the function on which the minimum is achieved. Substituting (5.1) and (5.2) into (5.3) we get

$$E_{min} = \min_{c_{\{n'\}}} \sum_{\{n'\}} \Big[\sum_{i=1}^{N} \sum_{r} \Big(e^{\frac{i}{N}(K,R)} \, \overline{c_{\{n'\}}} \, c_{\{(n' \setminus n_i)(n_i+r)\}} \cdot$$

$$J_{\{n'\}}^{\{(n' \setminus n_i)(n_i+r)\}} \Big) + \Big(\sum_{i \neq j} V_{n_i, n_j} \Big) \overline{c_{\{n'\}}} \, c_{\{n'\}} \Big] \tag{5.4}$$

If

$$J_{\{n'\}}^{\{(n' \setminus n_i)(n_i+r)\}} \leqslant 0 \ , \quad V_{n_i, n_j} \leqslant 0 \ , \tag{5.5}$$

then the minimum is achieved at K=0.

We find that the N-quasi-particle Hamiltonian (5.1) has under the conditions (5.5) a noncurrent ground state, i.e., the centre-of-mass quasi-momentum of the ground state equals to zero.

Let us now give an example of a one-dimensional two-quasi-particle system with a current ground state. Let $|k| \leqslant \frac{\pi}{2}$ and let the corresponding dispersion law be given by

$$e(k) = \begin{cases} 0 \ , & |k| < \frac{\pi}{4} \\ 1 \ , & \frac{\pi}{4} \leqslant |k| \leqslant \frac{\pi}{2} \end{cases} \tag{5.6}$$

Further let the two-particle interaction be given in the momentum representation by a rank-one operator with a kernel $V(q)V(q')$, where

$$V(q) = \begin{cases} 0 \ , & |q| < \frac{\pi}{4} \\ 1 \ , & \frac{\pi}{4} \leqslant |q| \leqslant \frac{\pi}{2} \end{cases} \tag{5.7}$$

Then the bound state energy E can be obtained from the equation (U is the coupling constant):

$$U \int_{\pi/4}^{\pi/2} \frac{V^2(q)\,dq}{e(\frac{K}{2}+q) + e(\frac{K}{2}-q) - E} = 1 \tag{5.8}$$

$$K=0: \quad 2U \int_{\pi/4}^{\pi/2} \frac{dq}{2-E} = \frac{\pi U}{2} \, \frac{1}{2-E} \ , \qquad E = 2 - \frac{\pi}{2}U$$

$$K=\frac{\pi}{2}: \quad 2U\int_{\pi/4}^{\pi/2}\frac{dq}{1/2-E}=\frac{\pi U}{2}\frac{1}{1/2-E}, \qquad E=\frac{1}{2}-\frac{\pi}{2}U$$

Hence $E(\frac{\pi}{2})<E(0)$. We note that in this example the inequalities $J_r\leq 0$, $V_r \gtrless 0$ hold.

The ground state energy of a N-quasi-particle system with a kinetic energy fulfiling min $e(k) = e(0)$ and with repulsive interaction between quasi-particles is given by $Ne(0)$.

References

1. D.Mattis, Rev.Mod.Phys. 58 (1986), 361-379
2. A.M.Kosevich, B.A.Ivanov, A.S.Kovalev,"Nonlinear magnetisation waves. Dynamical and topological solitons",Naukova Dumka, Kiev 1984 (in Russian)
3. R.A.Minlos, A.I.Mogilner, in "Mathematical problems of statistical mechanics and dynamics", D.Reidel, New York 1986
4. M.Reed, B.Simon,"Methods of modern mathematical physics. 4: Analysis of operators", Academic Press New York 1978
5. M.Reed, B.Simon,"Methods of modern mathematical physics. 3: Scattering Theory", Academic Press, New York 1979
6. A.I.Mogilner, VINITI Publications 1019-B87,1987 in Russian
7. H.Zolondek, Teor.Mat.Fis., 53 (1982), 216 - 226 (in Russian)
8. A.I.Mogilner, Physics of Metals and Metallography, in print (in Russian)
9. G.V.Efimov, Phys. Lett., B33 (1970), 563-564
10. D.R.Jafaev, Mat.Sbornik, 94 1974 , 567-593
11. R.D.Amado, J.V.Noble, Phys.Rev D5 1982 , 1992-2002
12. S.P.Merkurjev, L.D.Faddeev,"Quantum scattering theory for a few particle systems", Moscow, Nauka,(1985) (in Russian)
13. S.A.Vugalter, G.M.Zislin, Doklady Akademii Nauk SSSR, v.267 (1982), p.784 - 786 (in Russian)

CONTACT AND SURFACE PHENOMENA

SURFACES WITH AN INTERNAL STRUCTURE

B.S.Pavlov, P.B.Kurasov
Department of Mathematical and
Computational Physics, Institute for
Physics, Leningrad State University,
198904 Leningrad,St.Peterhoff, USSR

Zero-range potentials with internal structure allow to construct quantum-mechanical models of various physical phenomena. In this lecture, we discuss a particular class of operators of this type, which can be used for description of surface phenomena in solid-state phisics.

The concept of zero-range potentials with internal structure proposed in [1] opens way to construction of new exactly solvable quantum mechanical models. In the present paper this method is used to discuss the surface effects. The boundary-contact problem without internal structure was investigated rigorously earlier by B.P.Belinsky [2]. In distinction to the work of Karwowsky [3] who considered the simplest model of a translationally invariant surface, our method allows to treat the translationally invariant surfaces as well as periodic structures. Our approach is similar to the method of Ref. [4], where a similar mathematical construction was applied to the description of energy-dependent two-body problems in few-body physics.

The scattering on a surface, described in Sec.I is quite similar to the two-particle scattering. The spectrum of the corresponding operator is purely continuous. It consists of two branches : the branch of the scattered waves and the waveguide branch, which represents an analogue to bound states. The eigenfunctions corresponding to the first branch describe various processes of transmission and reflection of the plane waves. We consider here only the simplest case of a planar surface, but all the results can be generalised for an arbitrary shaped surface. Such a generalisation, however, does not yield an exactly solvable model.

The presence of a potential barrier (considered in Sec. 3) splits the scattering branch of the continuous spectrum into two parts.The surface waves described in Sec.2 can propagate with a velocity which differs from the velocity of the "external" waves. At the same time the waveguide branch of the spectrum changes. The space anisotropy of the surface modifies the waveguide functions in a typical way. Sec.4 is devoted to description of a model of a zero-range defect localised near the surface with an internal structure. In the last Section we discuss a surface with a periodical structure.

1. Formulation of the problem

Let Γ denote a plane in R^3,

$$\Gamma = \left\{ x=(x,y,z) \in R^3,\ z=0 \right\}$$

Let $A_0 = -\Delta$ be the self-adjoint Laplace operator in $L^2(R^3)$ with the standart domain $D(A_0)=W_2^2(R^3)$. We restrict this operator to $C_0^\infty(R^3\backslash\Gamma)$,

$$A_{00} = A_0 \upharpoonright C_0^\infty(R^3\backslash\Gamma)$$

where $C_0^\infty(\Omega)$ denotes the set of all infinitely differentiable functions with a compact support contained in Ω. This symmetric operator has infinite deficiency indices. The boundary form of the adjoint operator A_{00} restricted to the linear set of smooth functions can be written in terms of the limits of functions and their derivatives taken from the different sides of the plane Γ :

$$u_0^{\pm}(x,y) = u_0(x,y,\pm 0)$$

$$\frac{\partial u_0^{\pm}}{\partial n}(x,y) = \frac{\partial u_0^{\pm}}{\partial n}(x,y,\pm 0)$$

Then the boundary form can be expressed as follows :

$$\left\langle A_{00}^* u_0, v_0 \right\rangle - \left\langle u_0, A_{00}^* v_0 \right\rangle = \iint dx\, dy \left\{ \left[\frac{\partial u_0}{\partial n}\right] \overline{\langle v_0 \rangle} + \left\langle \frac{\partial u_0}{\partial n} \right\rangle \overline{[v_0]} - \right.$$

$$\left. - [u_0] \overline{\left\langle \frac{\partial v_0}{\partial n} \right\rangle} - \langle u_0 \rangle \overline{\left[\frac{\partial v_0}{\partial n}\right]} \right\} \equiv [u_0, v_0]^{\partial} \qquad \text{(I.I)}$$

where :

$$[u_0] = u_0^+ - u_0^- \qquad \left[\frac{\partial u_0}{\partial n}\right] = \frac{\partial u_0^+}{\partial n} - \frac{\partial u_0^-}{\partial n}$$

$$\langle u_0 \rangle = (u_0^+ + u_0^-)/2 \qquad \left\langle \frac{\partial u_0}{\partial n} \right\rangle = \frac{1}{2}\left(\frac{\partial u_0^+}{\partial n} + \frac{\partial u_0^-}{\partial n}\right)$$

In order to add an internal structure we choose self-adjoint operators A_i, $i = 1,2$ acting in Hilbert spaces H_i. In accordance with the standart procedure [1], we restrict these operators to symmetric operators with deficiency indices (1,1). An arbitrary element from the domain of the adjoint operator A_{i0}^* can be represented in the following form :

$$u_i = u_i^+ w_+^i + u_i^- w_-^i + \tilde{u}_i$$

where $\tilde{u}_i \in D(A_{i0})$ and $w_\pm^i$ are basis elements in the deficiency subspaces [1] . The boundary form of the adjoint operator can be described in terms of the "boundary" values u_i^+, u_i^- :

$$\langle A_{i0}^* u_i, v_i \rangle - \langle u_i, A_{i0}^* v_i \rangle = u_i^- \overline{v_i^+} - u_i^+ \overline{v_i^-} \equiv [u_i, v_i] \qquad (1.2)$$

Now we can construct symmetric operators with infinite deficiency indices. We start with self-adjoint operators defined in the Hilbert spaces $L^2(R^2, H_i)$ as follows :

$$(\mathcal{A}_i u)(x,y) = -\left(\frac{\partial^2 u_i}{\partial x^2} + \frac{\partial^2 u_i}{\partial x^2}\right)(x,y) + (A_i u_i)(x,y)$$

$$D(\mathcal{A}_i) = W_2^2(R^2, D(A_i))$$

We restrict these operators to the linear set of smooth functions $C^\infty(R^2, D(A_{i0}))$. The boundary form of the adjoint operator restricted to the set of smooth functions can be written as :

$$\langle \mathcal{A}_{i0}^* u_i, v_i \rangle - \langle u_i, \mathcal{A}_{i0}^* v_i \rangle = \iint dx\, dy\, (u_i^- \overline{v_i^+} - u_i^+ \overline{v_i^-}) \equiv [u_i, v_i]^\partial \qquad (1.3)$$

We shall consider the symmetric operator $\mathcal{L}_0$ acting in the Hilbert space

$$L_2(R^3) \oplus L_2(R^2, H_1) \oplus L_2(R^2, H_2)$$

whose elements are denoted by $U=(u_0,u_1,u_2)$ as follow

$$\mathcal{L}_0 = \mathcal{A}_{00} \oplus \mathcal{A}_{10} \oplus \mathcal{A}_{20}$$

This operator has infinite deficiency. The boundary form of its adjoint on the set of smooth functions represents a sum of boundary forms of the operators $\mathcal{A}_{00}, \mathcal{A}_{10}, \mathcal{A}_{20}$:

$$\langle \mathcal{L}_0^* U, V \rangle - \langle U, \mathcal{L}_0^* V \rangle = \sum_{\alpha=0}^{2} [u_\alpha, v_\alpha]^\partial$$

This form vanishes on the domain $D_0^{\alpha\beta}$ which is determined by the boundary conditions :

$$\left[\frac{\partial u_0}{\partial n}\right] = \alpha u_1^- \qquad [u_0] = \beta u_2^+ \tag{1.4}$$
$$u_1^+ = \bar{\alpha} \langle u_0 \rangle \qquad u_2^- = \bar{\beta} \left\langle \frac{\partial u_0}{\partial n} \right\rangle$$

Thus the restriction of the adjoint operator $\mathcal{L}_0^*$ to the domain $D_0^{\alpha\beta}$ is a below semibounded symmetric operator. We can therefore construct Friedrichs extention $\mathcal{L}^{\alpha\beta}$ of the symmetric operator $\mathcal{L}_0$. One can show that $\mathcal{L}^{\alpha\beta}$ does not coincide with the original operator so this Friedrichs extension is not trivial. The spectrum of this operator is purely continuous. There exist two branches of the continuous spectrum : the scattering branch and the waveguide branch. The external component of the scattered waves can be decomposed into plain waves of three types :

incoming waves $\exp i(k,x)$
reflected waves $R \exp i(\tau k,x)$
transmitted waves $S \exp i(k,x)$

(see Fig.1)

where :

$$\tau(k_x,k_y,k_z) = (k_x,k_y,-k_z)$$

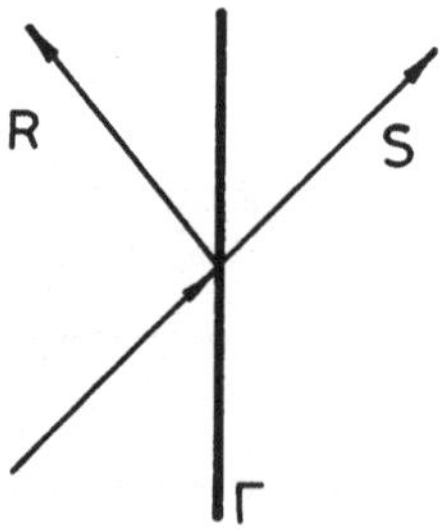

Fig. 1

The external component satisfies the following equation :

$$-\Delta u_0 = k^2 u_0$$

The boundary values for the external component are :

$$\langle u_0 \rangle = \frac{1}{2} (1 + S + R) \exp i(k_x x + k_y y)$$

$$[u_0] = (S - 1 - R) \exp i(k_x x + k_y y)$$

$$\langle \frac{\partial u_0}{\partial n} \rangle = \frac{ik_z}{2} (1 + S - R) \exp i(k_x x + k_y y)$$

$$\left[\frac{\partial u_0}{\partial n} \right] = ik_z (S - 1 + R) \exp i(k_x x + k_y y) \qquad (1.5)$$

The internal components are assumed to be of the following form :

$$u_i(x,y) = f(x,y)\, u_i$$

where $f \in W^2_{2,loc}(R^2)$, $u_i \in H_i$. Let $f(x,y)$ be given by

$$f(x,y) = \exp i(k_x x + k_y y)$$

We obtain the following equations for elements of the internal spaces H_i :

$$A_{i0} u_i = (K^2 - k_x^2 - k_y^2) u_i = k_z^2 u_i$$

It follows [1] that

$$\frac{u_i^-}{u_i^+} = D_i(k_z^2)$$

where D_i are Schwarz integrals of the internal operators A_i on the deficiency elements θ_i :

$$D_i(\ ) = \left\langle \frac{I + \lambda A_i}{A_i - \lambda} \theta_i, \theta_i \right\rangle$$

Substituting $u_0^{\pm}$, $u_i^{\pm}$ into the boundary conditions (1.4), we can calculate the transmission and reflection amplitudes :

$$S = \frac{\left(\frac{|\alpha|^2}{2D_1} - \frac{2}{|\beta|^2 D_2}\right) ik_z}{\left(ik_z - \frac{|\alpha|^2}{2D_1}\right)\left(ik_z - \frac{2}{|\beta|^2 D_2}\right)}$$

$$R = \frac{(ik_z)^2 - \frac{|\alpha|^2}{2D_1} \frac{2}{|\beta|^2 D_2}}{(ik_z - \frac{|\alpha|^2}{2D_1})(ik_z - \frac{2}{|\beta|^2 D_2})} \qquad (1.6)$$

They fulfil the identity :

$$|R|^2 + |S|^2 = 1$$

The internal components u_1, u_2 of U can be easily calculated in terms of the deficiency elements θ_i.

Let us investigate now behaviour of the scattering matrics in the case of a weak coupling between the internal and external channels, $\alpha \to 0$ and $\beta \to 0$. The transmission amplitude S vanishes if

$$\frac{|\alpha|^2}{2D_1} - \frac{2}{|\beta|^2 D_2} = 0$$

The zero lines of this equation in the (k,t) - plane coincide (approximately) with the zero lines of the function D_1/D_2 . The reflection amplitude R vanishes if

$$D_1 D_2 = \frac{|\alpha|^2}{|\beta|^2} \frac{1}{(ik_z)^2}$$

The eigenfunction of this type form the $[0,\infty)$ branch of the continuous spectrum.

The external component of waveguide functions can be found in the following way : we put

$$u_0(x,y,z) = \begin{cases} B^+ \exp(-\varkappa z) & \exp i(k_x x + k_y y),\ z>0\ ,\ \varkappa>0 \\ B^- \exp(\varkappa z) & \exp i(k_x x + k_y y),\ z<0, \varkappa>0 \end{cases} \quad (1.8)$$

So the boundary values are

$$\left[\frac{\partial u_0}{\partial n}\right] = -\varkappa(B^+ + B^-) \quad \exp i(K_x x + k_y y)$$

$$\left\langle \frac{\partial u_0}{\partial n} \right\rangle = \frac{\varkappa}{2} (B^- - B^+) \exp i(k_x x + k_y y)$$

$$\left[u_0 \right] = (B^+ - B^-) \exp i(k_x x + k_y y)$$

$$\left\langle u_0 \right\rangle = \frac{1}{2} (B^+ + B^-) \exp i(k_x x + k_y y)$$

The conditions (1.4) now yield

$$\left[\frac{\partial u_0}{\partial n} \right] = \frac{|\alpha|^2}{D_1(-\varkappa^2)} \quad \left\langle u_0 \right\rangle$$

$$\left[u_0 \right] = |\beta|^2 D_2(-\varkappa^2) \left\langle \frac{\partial u_0}{\partial n} \right\rangle \tag{1.9}$$

The constants B^+, B^- satisfy the following homogeneous linear system of equations

$$(\varkappa + \frac{|\alpha|^2}{2 D_1}) B^+ + (\varkappa + \frac{|\alpha|^2}{2 D_1}) B^- = 0$$

$$(1 + \varkappa \frac{|\beta|^2 D_2}{2}) B^+ - (1 + \varkappa \frac{|\beta|^2 D_2}{2}) B^- = 0 \tag{1.10}$$

which has a nontrivial solution only in the following cases

$$\varkappa + \frac{|\alpha|^2}{2 D_1} = 0$$

$$1 + \varkappa \frac{|\beta|^2 D_2}{2} = 0 \tag{1.11}$$

The solution of the first equation in (1.11) gives symmetric waveguide functions while the solution of the second equation gives the antisymmetric ones. In the weak-coupling case the solutions of the equations (1.11) are near to the zero lines of D_1 and D_2^{-1}, respectively. The waveguide eigenfunctions correspond to the following branches of the continuous spectrum :

$$[-\varkappa_i^2, \infty) \quad ,$$

where $\varkappa_i$ are arbitrary real solutions of one of the equations (1.11).

A screen with an internal structure allows us to construct a resonance transmission of plane waves across the surface. The external components S_{oo} of the scattering matrix depends on the normal component of the momentum k_z. In this way we can obtain waveguide eigenfunctions in this surface model.

2. Velocity of waves in the internal space

In this section we are going to describe a model of the surface with a special internal channel. The velocity of internal waves is, in general, different from the velocity of the external ones. Let $\mathcal{A}_i$ be the unperturbed internal operator :

$$\mathcal{A}_i = - c_i^2 \Delta_{xy} + A_i$$

The constants $c_i \in R_+$ are velocities of the internal waves. One can construct a self-adjoint operator following the procedure outlined in the previous section.

The spectrum of such an operator is also purely continuous and all eigenfunctions can be calculated in the same way as in the Sec.1. The waveguide functions depends on the quasi-momentum $t=(t_x,t_y)$ and the same is true for the energy λ. One can easily obtain the dispersion equations :

$$-(\lambda - t^2) + \frac{|\alpha|^2}{2\, D_1(\lambda - c_1^2 t^2)} = 0 \tag{2.1}$$

$$-1 + (\lambda - t^2)\, \frac{|\beta|^2 D_2(\lambda - c_1^2 t^2)}{2} = 0 \tag{2.2}$$

Let us investigate the solution of these equations in the case of weak coupling between the external and internal channels. There are two different cases to be distinguished

1) $c_i > 1$
2) $c_i < 1$

In the first case all energy bands are finite while in the second case they are infinite. We shall prove this result graphically using Fig.2 . Zero lines of the Schwarz integral $D_1(-c_1^2 t^2)$ are parabolas $\lambda - c_1^2 t^2 = \text{const}$ denoted by the dotted lines. In the case $c_1 > 1$, the parabolas of this family intersect the parabola $\lambda = t^2$ or lie inside it (Fig.2a). If $c_1 < 1$ then the parabolas lie in the exterior of the parabola $\lambda = t^2$ or intersect it (Fig.2b).

Solutions to the equations (2.1), (2.2) are situated near the zero lines of the Schwarz integrals in the region $\lambda < t^2$. They are denoted by full lines on the figures. The projection of the solutions on the energy axis yield the spectral bands of the corresponding operator.

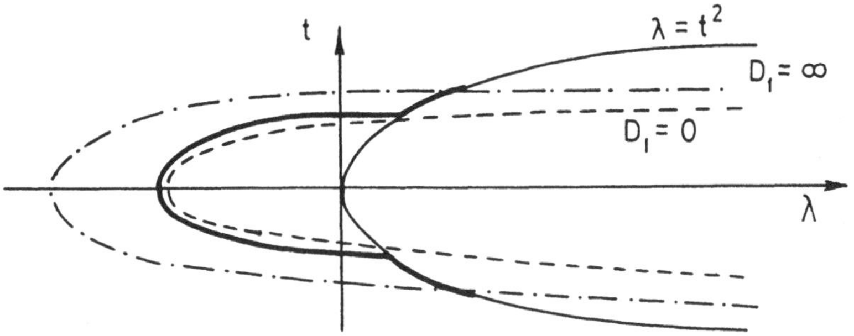

Fig.2a

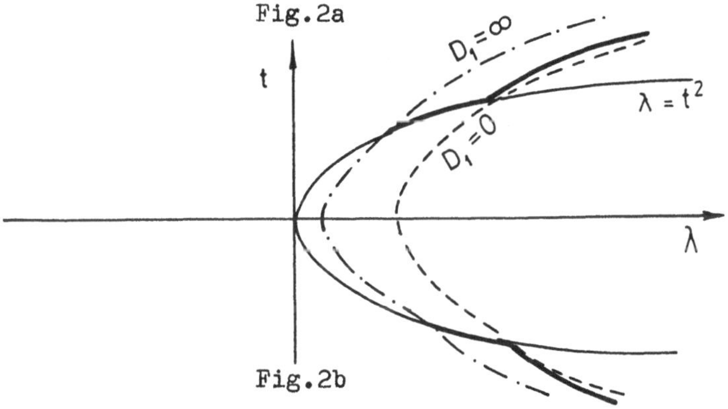

Fig.2b

The simplest generalisation of this model is connected with the space anisotropy of the surface. Suppose that the unperturbed internal operator $\mathcal{A}_i$ is of the form :

$$\mathcal{A}_i = - c_{ix}^2 \frac{\partial^2}{\partial x^2} - c_{iy}^2 \frac{\partial^2}{\partial x^2} + A_i$$

In the case $c_{ix}=c_{iy}$, the diagrams on Fig.2 become three-dimensional. Solutions to the corresponding equations lie near the zero surfaces of the functions $D_1(\lambda - c_{1x}^2 t_x^2 - c_{1y}^2 t_y^2)$ and

$$D_2^{-1}(\lambda - c_{2x}^2 t_x^2 - c_{2y}^2 t_y^2)$$

The picture is symmetric with respect to the energy axis. The wave-guide functions can propagate nearly in a particular direction in one of the following cases :

$$c_{ix} \gg c_{iy} \qquad \text{or} \qquad c_{ix} \ll c_{iy}$$

The corresponding energy bands are infinite if in some direction the velocity of the internal wave is less than the velocity of the external one .

3. Surfaces with a potential barrier

Suppose now that the external unperturbed operator is of the form $\mathcal{A}_0 = -\Delta + q(z)$, where

$$q(z) = \begin{cases} 0, & z > 0 \\ -A, & z < 0 \end{cases}$$

For some $A > 0$ we can construct the internal structure in the same way as in Sec.1. The model is now nonsymmetric with respect to the plane Γ. Consequently the scattering matrices are different in the positive and negative z-axis directions, (S^+,R^+) and (S^-,R^-) respectively.

The external component of the scattered waves have the same form, as in the previous cases but the momenta of the incoming and transmitted waves are different. If we have an incoming wave of a momentum $k' = (k'_x, k'_y, k'_z)$, $k'_z > 0$ the reflected wave has the momentum $\tau k'$ and the transmitted one $k = (k'_x, k'_y, k'^2_z - A)$. If $k'_z - A < 0$, then the transmitted wave is damped. The standard procedure gives us the following results :

$$S^+ = \frac{2\, i k'_z \left(\dfrac{|\alpha|^2}{2\,D_1} - \dfrac{2}{|\beta|^2 D_2} \right)}{\left(ik_z - \dfrac{|\alpha|^2}{2\,D_1}\right)\left(ik'_z - \dfrac{2}{|\beta|^2 D_2}\right) + \left(ik'_z - \dfrac{|\alpha|^2}{2\,D_1}\right)\left(ik_z - \dfrac{2}{|\beta|^2 D_2}\right)}$$

$$R^+ = \frac{\left(ik_z - \dfrac{|\alpha|^2}{2\,D_1}\right)\left(ik'_z + \dfrac{2}{|\beta|^2 D_2}\right) + \left(ik'_z + \dfrac{|\alpha|^2}{2\,D_1}\right)\left(ik_z - \dfrac{2}{|\beta|^2 D_2}\right)}{\left(ik_z - \dfrac{|\alpha|^2}{2\,D_1}\right)\left(ik'_z - \dfrac{2}{|\beta|^2 D_2}\right) + \left(ik'_z - \dfrac{|\alpha|^2}{2\,D_1}\right)\left(ik_z - \dfrac{2}{|\beta|^2 D_2}\right)}$$

One can check directly that $R^+ = 1$ holds for $k_z' < A$; in this case the incoming wave is totally reflected by the potential barrier.

The eigenfunctions of this type form the waveguide branch $[-A,\infty)$ of the continuous spectrum.

The scattered waves corresponding to the normal component of the incoming wave $k_z < 0$ can be described in the same way :

$$S^- = \frac{2\, ik_z \left(\frac{2}{|\beta|^2 D_2} - \frac{|\alpha|^2}{2\, D_1} \right)}{\left(ik_z' + \frac{|\alpha|^2}{2\, D_1}\right)\left(ik_z + \frac{2}{|\beta|^2 D_2}\right) + \left(ik_z + \frac{|\alpha|^2}{2\, D_1}\right)\left(ik_z' + \frac{2}{|\beta|^2 D_2}\right)}$$

$$R^- = \frac{\left(ik_z' + \frac{|\alpha|^2}{2\, D_1}\right)\left(ik_z - \frac{2}{|\beta|^2 D_2}\right) + \left(ik_z - \frac{|\alpha|^2}{2\, D_1}\right)\left(ik_z' + \frac{2}{|\beta|^2 D_2}\right)}{\left(ik_z' + \frac{|\alpha|^2}{2\, D_1}\right)\left(ik_z + \frac{2}{|\beta|^2 D_2}\right) + \left(ik_z + \frac{|\alpha|^2}{2\, D_1}\right)\left(ik_z' + \frac{2}{|\beta|^2 D_2}\right)}$$

These eigenfunctions form the branch $[0,\infty)$ of the continuous spectrum.

If $\frac{|\alpha|^2}{2\, D_1} - \frac{2}{|\beta|^2 D_2} = 0$ then the transmission amplitudes $S^\pm$ are equal to zero. while the moduli of the corresponding reflection amplitudes are equal to one.

The waveguide functions can be constructed in the standard form

$$u_0(x,y,z) = \begin{cases} B^+ \ \exp(-\varkappa z)\, \exp i(k_x x + k_y y)\ , \ z > 0 & \varkappa, \varkappa' > 0 \\ B^- \ \exp(+\varkappa' z)\, \exp i(k_x x + k_y y)\ , \ z < 0 & \varkappa' = \sqrt{\varkappa^2 - A} \end{cases}$$

One can easily obtain the following dispersion equations :

$$\frac{|\beta|^2 D_2}{2} = - \frac{\varkappa + \varkappa' + 2\, \frac{|\alpha|^2}{2\, D_1}}{2\varkappa\varkappa' + (\varkappa + \varkappa')\, \frac{|\alpha|^2}{2\, D_1}}$$

$$-\frac{|\alpha|^2}{2D_1} = \frac{\varkappa+\varkappa'+2\varkappa\varkappa'\frac{|\beta|^2 D_2}{2}}{2+(\varkappa+\varkappa')\frac{|\beta|^2 D_2}{2}}$$

Thus the potential barrier splits the scattering branch of the spectrum into two parts $[-A,\infty)$ and $[0,\infty)$ and changing at the same time the waveguide branch.

4. A zero-range defect near the surface

Let us restrict the self-adjoint operator $\mathcal{L}^{\alpha\beta}$ of the section 1 to the linear set of smooth functions vanishing in a neighbourhood of a point $(0,0,-z_0)$, $z_0 > 0$. The obtained symmetric operator $\mathcal{L}^{\alpha\beta}_{z_0}$ has the deficiency indices (1,1). The external component of an element from the domain of the adjoint operator $(\mathcal{L}^{\alpha\beta}_{z_0})^*$ has the following asymptotic behaviour in a neighbourhood of the point z_0 :

$$u_0(x,y,z) \sim \frac{u_0^+}{4\pi R_1} + u_0^- + o(R_1)$$

where $R_1 = (x,y,z+z_0)$. The boundary form of the adjoint operator can be written in the standard way

$$\langle \mathcal{L}^{\alpha\beta *}_{z_0} U, V \rangle - \langle U, \mathcal{L}^{\alpha\beta *}_{z_0} V \rangle = u_0^- \overline{v_0^+} - u_0^+ \overline{v_0^-} = [U,V]^{ex} \quad (4.1)$$

The deficiency element $G(k)$ corresponding to the spectral parameter $k=\sqrt{\lambda}$, $\mathrm{Im}\, k > 0$, has the external component of the following type:

$$G_0 = \begin{cases} \dfrac{\exp ikR_1}{4\pi R_1} + \widetilde{G}_0 , & z < 0 \\ \widetilde{G}_0 , & z > 0 \end{cases}$$

where $\widetilde{G}_0$ is the result of scattering of the spherical wave $\dfrac{\exp ikR_1}{4\pi R_1}$ on the surface with the internal structure. The unknown function G_0 can be calculated with the help of the following expansion (see[5])

$$\frac{\exp ikR_1}{4\pi R_1} = \frac{ik}{8\pi^2}\int_0^{\pi/2 - i\infty} d\theta \int_0^{2\pi} d\varphi \; \exp i(k_x x + k_y y + k_z(z+z_0)) \sin\theta, \quad z \geq z_0$$

where $k_x = k \sin\theta \cos\varphi$ $\qquad k_z = k \cos\theta$

$k_y = k \sin\theta \cos\varphi$

Plane waves coming to the surface stimulate reflected and transmitted waves. The formulas (1.6) are valid in the case of damped waves too, because all the functions involved are analytic in the upper semiplane. For $\widetilde{G}_0$ we get the following expressions :

the reflected wave

$$\widetilde{G}_0 = \frac{ik}{8\pi^2} \int d\theta \int d\varphi \ \exp i(k_x x + k_y y - k_z(z-z_0)) \ R(k\cos\theta) \sin\theta$$

the scattered wave

$$\widetilde{G}_0 = \frac{ik}{8\pi^2} \int d\theta \int d\varphi \ \exp i(k_x x + k_y y + k_z(z+z_0)) \ S(k\cos\theta) \sin\theta$$

where S and R are the transmission and reflection amplitudes introduced in Sec.1. Symmetry of the model allows us to transform these expressions to

$$\widetilde{G}_0 = \frac{ik}{8\pi} \int_{-\pi/2+i\infty}^{\pi/2-i\infty} H_0^{(1)}(kR' \sin\theta) \exp(-ik(z-z_0)\cos\theta) R(k\cos\theta)\sin\theta \, d\theta \qquad R' = |(x,y,z-z_0)|$$

and

$$G_0 = \frac{ik}{8\pi} \int H_0^{(1)}(kR_1 \sin\theta) \exp(ik(z+z_0)\cos\theta) \ S(k\cos\theta) \sin\theta \, d\theta$$

The internal components can be easily calculated in the same way. The boundary values G_0^+, G_0^- for the deficiency element are the following :

$$G_0^+ = 1 , \quad G_0^- = \frac{ik}{4\pi}\left(1 + \int_0^{\pi/2-i\infty} \exp(ik\cos\theta \, 2z_0) \ R(k\cos\theta) \sin\theta \, d\theta \right)$$

The asymptotic behaviour at infinity can be investigated by the method of Brechovskich [5] . If we fix a direction specified by an angle θ, then

$$\widetilde{G}_0 \sim \frac{\exp ikR_1}{4\pi R_1} \ R(k\cos\theta) , \quad \theta \in \left[-\frac{\pi}{2}, \frac{\pi}{2}\right]$$

$$\widetilde{G}_0 \sim \frac{\exp ikR_1}{4\pi R_1} \ S(k\cos\theta) , \quad \theta \in \left[\frac{\pi}{2}, \frac{3}{2}\pi\right]$$

This deficiency element is the kernel for the external component of the resolvent of the self-adjoint operator described in Sec.1 .

In order to construct the zero-range potential with an internal structure, let us choose a self-adjoint operator A_3 acting in the Hilbert space H_3. We shall restrict it to a symmetric operator A_0 with deficiency indices (1,1). The boundary form can be expressed in a standard way :

$$\langle A_{30}^{*} u_3, v_3 \rangle - \langle u_3, A_{30}^{*} v_3 \rangle = u_3^{-} \overline{v_3^{+}} - u_3^{+} \overline{v_3^{-}} \equiv [u_3, v_3]^{in} \quad (4.2)$$

Next we define the symmetric operator $N_0 = \mathcal{L}_{z_0}^{\alpha\beta} \oplus A_{30}$ acting in the Hilbert space $L_2(R^3) \oplus L_2(R^2, H_1) \oplus L_2(R^2, H_2) \oplus H_3$. Its deficiency indices are (2.2), boundary form is the sum of the boundary forms :

$$\langle N_0^{*} U, V \rangle - \langle U, N_0^{*} V \rangle = [U, V]^{ex} + [u_3, v_3]^{in}$$

We can define the sought self-adjoint operator by restricting the adjoint operator N_0^{*} to the linear set specified by the boundary conditions :

$$u_0^{+} = \gamma\, u_3^{+}$$
$$-u_3^{-} = \overline{\gamma}\, u_0^{-}$$

The continuous spectrum of N_γ consists of eigenfunctions of two types, the scattered waves and waveguide functions. The external components of all eigenfunctions of the continuous spectrum can be expressed in the following form :

$$u_0^{N}(\lambda) = u_0^{\mathcal{L}}(\lambda) + \rho\, G_0(\lambda)$$

where $u_0^{\mathcal{L}}(\lambda)$ are the eigenfunctions of the self-adjoint operator $\mathcal{L}^{\alpha\beta}$. The boundary values of such function are the following :

1) $u_0^{\mathcal{L}+} = 0$ $\quad u_0^{\mathcal{L}-} = \exp{-ik_z z_0} + R \exp{ik_z z_0}\,,\ k_z > 0$

$\quad u_0^{\mathcal{L}-} = S \exp{-ik_z z_0}\,,\ k_z < 0$

2) $u_0^{\mathcal{L}+} = 0$ $\quad u_0^{\mathcal{L}-} = B^{-} \exp \varkappa z_0$

Hence we can calculate the amplitude ρ from the boundary conditions and from the standard correlation between the boundary values of the internal compenent

$$\frac{u_3^{-}}{u_3^{+}} = D_3(\lambda)$$

where D_3 is the Schwarz integral of the operator A_3 .

The amplitude ϱ is expressed as follows

$$\varrho = \frac{-u_0^{\alpha -}}{|\gamma|^{-2} D_3(\lambda)\, G_0^+ + G_0^-}$$

The external components of the scattered waves has the following asymptotics at infinity

$$u_0^N(\lambda) \sim u_0^{\alpha}(\lambda) + f(\vec{k},\theta) \frac{\exp ikR_1}{4 \ R_1}$$

where

$$f(k,\theta) = \varrho\, R(k \cos\theta), \quad \theta \in \left]\frac{\pi}{2}, \frac{3}{2}\pi\right]$$

$$f(k,\theta) = S(k \cos\theta), \quad \theta \in \left[-\frac{\pi}{2}, \frac{\pi}{2}\right]$$

and the function $f(\vec{k},\theta)$ is the scattering amplitude.

The eigenvalues of the discrete spectrum correspond to solutions of the equation :

$$\frac{G_0^-}{G_0^+} = - \frac{D_3(\lambda)}{|\gamma|^2}$$

In the case of weak coupling, $\gamma \to 0$, the eigenvalues are situated near the negative zeroes of $D_3(\lambda)$, while all positive solutions of this equation correspond to resonances.

5. A periodic structure on the surface

There are different ways how to construct such a periodic structure within our approach. One can choose the internal operators A_i dependent on the point (x,y) of the surface in some periodic way. One can add also a periodic potential $g(x,y)$ into the definition of the operator $\mathcal{A}_i$:

$$\mathcal{A}_i = -\Delta_{xy} + A_i + g(x,y)$$

We are going to investigate here the simplest exactly solvable model. We can replace the constants α and β in the boundary conditions (1.4) by arbitrary periodic smooth functions $a(x,y)$ and $b(x,y)$. Let a, b be the following exponential functions

$$a(x,y) = \alpha \ \exp i(p_{1x}x + p_{1y}y)$$

$$b(x,y) = \beta \ \exp i(p_{2x}x + p_{2y}y)$$

Let us investigate the spectrum of the waveguide functions. The external components constructed in a standard way :

$$u_0(x,y,z) = \begin{cases} B^+ \exp -\varkappa z \quad \exp i(t_x x + t_y y) , & z > 0 \\ B^- \exp \varkappa z \quad \exp i(t_x x + t_y y) , & z < 0 \end{cases}$$

From the boundary conditions we get that the constants B^+, B^- solve the standard homogeneous system :

$$\left(\varkappa + \frac{|\alpha|^2}{2\, D_1(\lambda - (t - p_1)^2)}\right) B^+ + \left(\varkappa + \frac{|\alpha|^2}{2\, D_1(\lambda -(t-p_1)^2)}\right) B^- = 0$$

$$\left(1 + \varkappa \frac{|\beta|^2 D_2(\lambda -(t-p_2)^2)}{2}\right) B^+ - \left(1+ \varkappa\frac{|\beta|^2 D_2(\lambda -(t-p_2)^2)}{2}\right) B^- = 0$$

The system has a nontrivial solution only if

$$\varkappa + \frac{|\alpha|^2}{2\, D_1(\lambda -(t-p_1)^2)} = 0$$

$$1 + \varkappa \frac{|\beta|^2 D_2(\lambda -(t-p_2)^2)}{2} = 0$$

Let us investigate the solutions of the first equation which lead to symmetric wavefunctions. The real solutions can be localised only in the region $\lambda < t^2$ (see Fig.3) because $\varkappa = \sqrt{t^2 - \lambda}$. In the case of weak coupling between the external and internal channels, the solutions lie near the zero surfaces of $D_1(\lambda -(t-p)^2)$ (marked by the

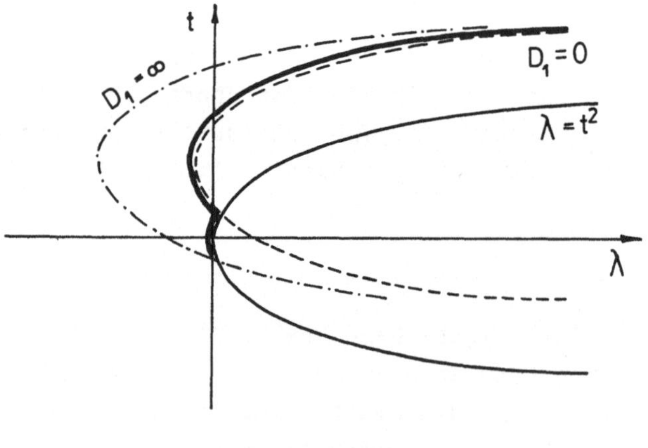

Fig.3

dashed lines), which are parabolas of the following type :

$$\lambda - (t - p)^2 = \text{const}$$

These parabolas are symmetric with respect to the axis t=p. Thus the branches of the solutions are infinite (marked by the full line), and they are not symmetric with respect to the energy axis.

If a waveguide function propagates in some direction, there is no waveguide function of the same energy propagating in the opposite direction. The same is true for the antisymmetric functions.

References

1. B.S.Pavlov, Teor. Mat. Fiz. 59 (1984), 345-353
2. B.P.Belinsky, Proc. Acad. Sci. USSR 278 (1984), 1090-1094
3. W.Karwowsky, in this volume
4. Yu.A.Kuperin,... : Properties of few-body quark-hadron systems, part II, Vilnius 1986, pp.28-73 (in Russian)
5. L.M.Brekhovskikh : Waves in sandwiched media, Nauka, Moscow 1973 (in Russian)

SPECTRAL PROPERTIES OF THE LAPLACIAN WITH ATTRACTIVE BOUNDARY CONDITIONS

Manfred Schröder, Sektion Mathematik, Karl-Marx-Universität, Karl-Marx-Platz 4-10, Leipzig 7010, German Democratic Republic

0. Introduction

Let a quantum particle be enclosed in a container G with the boundary ∂G. The dynamics of the particle can be described by the Hamiltonian

$$H = H_o + V, \tag{1}$$

where H_o is the Hamiltonian of a particle in a container with indifferent boundary: In an appropriate system of units, one has $H_o = -\Delta$ on $L_2(G)$ with Neumann boundary conditions (b.c.), and V is the potential corresponding to the boundary forces.

It can be shown /13/ that if G is the half-space $\mathbb{R}^d_+ = \mathbb{R}^{d-1} \times \mathbb{R}_+$, and V has the property that the function $Q(x_1,\ldots,x_{d-1}) = \int_0^\infty V(x_1,\ldots,x_d)dx_d$ belongs uniformly locally to $L_p(\mathbb{R}^{d-1})$, where $p > d-1$ and $p \geq 2$, then the sequence $H_o + nV(x_1,\ldots,x_{d-1},nx_d)$ converges for $n \to \infty$ inthe norm-resolvent sense to the operator $H_Q = -\Delta$ on $L_2(\mathbb{R}^d_+)$ with b.c.

$$\frac{\partial u}{\partial x_d}\bigg|_{x_d=0} = Qu\bigg|_{x_d=0} \tag{2}$$

(for d = 1, cf. Ref. 1). Since the boundary forces are usually of short range, it seems to be senseful to consider the Hamiltonian H_Q instead of (1).

If the function Q is non-negative, then the spectrum of H_Q coincides with that of H_o, even if we add to both Hamiltonians some external potential W. Such b.c. we call repulsive. Thereas if there are sufficiently large regions, where Q becomes negative, then there occur additional states - surface states - with wave functions concentrated near the boundary. In this case we call the b.c. attractive.

In the following we will consider three typical situations: In Section 1, we will investigate the effect of local impurities of a neutral surface. Section 2 is dedicated to ergodic surface potentials which can be caused e.g. by the molecular forces at the surface of a crystalline or amorphous solid. In the last section, we consider surface potentials varying remarkably only in a macroscopic scale.

1. Localized surface potentials

If the dynamics of a quantum particle in a container G is described by the Hamiltonian $H = H_Q + W$, where Q corresponds to the surface forces and W is an external potential, then the spectrum of H can be found by investigating the solvability of the equation $Hu = Eu$ for any real E. Generally, this requires methods described in /2/. In a quite simple way it can be done if $W = 0$ and $G = \mathbb{R}^d_+$ (cf. /3/). One obtains the following

Proposition 1: Let Q be relatively bounded with respect to $(-\Delta)^{1/2}$ with a relative bound less than 1. Then $0 > E \in \sigma(H_Q)(\sigma_{pp}(H_Q), \sigma_{ess}(H_Q))$ if and only if $0 \in \sigma(K_{Q,E})(\sigma_{pp}(K_{Q,E}), \sigma_{ess}(K_{Q,E}))$, where $\sigma(\sigma_{pp}, \sigma_{ess})$ denote the spectrum (pure point spectrum, essential spectrum), and $K_{Q,E} = (-\Delta - E)^{1/2} + Q$.

If the condition on Q is satisfied then $\sigma(H_Q)$ is bounded from below /6/. A sufficient condition on Q to be fit for Proposition 1 is that $Q \in L_p$ uniformly locally, with $p > d-1$, $p \geq 2$ /3/. We begin with the case $Q \in L_p$. This implies that $\sigma_{ess}(K_{Q,E}) = \sigma(K_{0,E})$, hence, by Proposition 1, $\sigma_{ess}(H_Q) = [0,\infty)$. Now let us denote by N(A,t) the number of eigenvalues (counting multiplicity) of the self-adjoint operator A, which do not exceed the real number t. From Proposition 1 it follows that, for $E < 0$, $N(H_Q,E) = N(K_{Q,E},0)$.

For $d = 2$, one can estimate $N(K_{Q,E},0)$ by using the Birman-Schwinger principle (cf. /4,11/).

Proposition 2: Let $Q \in L_2(\mathbb{R})$ and $Q_- \in L_2(\mathbb{R}) \cap L_p(\mathbb{R})$, $1 < p \leq 2$, and let $\frac{1}{p} + \frac{1}{q} = 1$. Then there exists a constant c_p such that for $E < 0$

$$N(H_Q,E) \leq c_p \, \|Q_-\|_p^2 \, (-E)^{-1/q}. \tag{3}$$

(Here and furtheron we denote by $F_\pm$ the positive (negative) part of F: $F_\pm = (|F| \mp F)/2$.)

In higher dimensions, one can use an analogy of the Cwikel-Lieb-Rosenbljum bound /10,11/ for the same purpose: For $Q \in L_p(\mathbb{R}^{d-1})$, with $d > 2$ and $p > d-1$, we set

$$\Phi_Q(q) = \text{mes}\left\{ x \in \mathbb{R}^{d-1} : Q(x) \leq q \right\}, \tag{4}$$

where mes denotes the Lebesgue measure. Now we state (cf. /6/)

Proposition 3: Let $Q \in L_p(\mathbb{R}^{d-1})$, with $d > 2$ and $p > d-1$. Then there are two constants $c_{p,d}$ and $c'_{p,d}$, such that, for $E < 0$,

$$N(H_Q,E) \leq \int_{-\infty}^{0} \min\left\{ c_{p,d}\,(-q)^{d-1},\; c'_{p,d}\,(-E)^{-1/2}\,(-q)^d \right\} d\Phi_Q(q). \tag{5}$$

For $Q_- \in L_{d-1} \cap L_d$, Proposition 3 implies $N(H_Q,E) \leq c_{p,d} \|Q_-\|_{d-1}^{d-1}$ and $N(H_Q,E) \leq c'_{p,d} \|Q_-\|_d^d \, (-E)^{-1/2}$.

2. Ergodic surface potentials

Now we suppose that Q^ω is some random ergodic function in the sense that any measurable set of realizations which is invariant under translations $Q^\omega(x) \mapsto Q^\omega(x+y)$, where y runs over $\mathbb{R}^{d-1}$ or some lattice $\mathbb{Z}_a^{d-1} = (a_1\mathbb{Z}) \times \ldots \times (a_{d-1}\mathbb{Z})$ with $a_i > 0$, has probability 0 or 1.

For given $L > 0$, let $\Lambda_L = [-L/2, L/2]^{d-1}$, and denote by $H^{\omega}_{Q,L}$ resp. $\tilde{H}^{\omega}_{Q,L}$ the restriction of H^{ω}_{Q} onto $\Lambda_L \times [0,L]$ resp. $\Lambda_L \times \mathbb{R}_+$, with appropriate (e.g. Neumann) b.c. at $\partial\Lambda_L \times [0,L]$ and $\Lambda_L \times \{L\}$ resp. at $\partial\Lambda_L \times \mathbb{R}_+$, and let analogously $K^{\omega}_{Q,E,L}$ denote the restriction of $K^{\omega}_{Q,E}$ onto Λ_L with the same b.c. The estimates of Section 1 suggest that it should be senseful to consider the quantity

$$n^{\omega}_L(E) = N(H^{\omega}_{Q,L}, E)/L^{d-1} \tag{6}$$

in the limit $L \to \infty$. It turns out that the quantity

$$\tilde{n}^{\omega}_L(E) = N(\tilde{H}^{\omega}_{Q,L}, E)/L^{d-1} = N(K^{\omega}_{Q,E,L}, 0)/L^{d-1} \tag{7}$$

(cf. /5/) has the same limit as $n^{\omega}_L(E)$. If this limit exists we call it the density of surface states. But when does it exist? One possible answer is the following /5/

Proposition 4: Let $Q^{\omega} \in L_p(\mathbb{R}^{d-1})$ uniformly locally, with $p \geq 2$, $p > d-1$, and let Q^{ω} be ergodic with respect to translations in $\mathbb{R}^{d-1}$ resp. $\mathbb{Z}^{d-1}_a$. If $\mathrm{Exp}(|Q^{\omega}(0)|^p) < \infty$ resp.

$$\mathrm{Exp}\left(\int_0^{a_1} dx_1 \ldots \int_0^{a_{d-1}} dx_{d-1} |Q^{\omega}(x)|^p\right) < \infty$$

then there exists a non-negative, monotonously increasing function $n^s(E)$ on $\mathbb{R}_-$, such that for all negative rational E almost surely

$$\lim_{L\to\infty} n^{\omega}_L(E) = \lim_{L\to\infty} \tilde{n}^{\omega}_L(E) = n^s(E).$$

This result is the analogy to the non-randomness of the density of states of Schrödinger operators with ergodic potentials (cf. e.g. /7,8/ and references therein).

The density of surface states can be estimated analogously as in the previous section: Let

$$\Psi_Q(q) = \lim_{L\to\infty} \text{mes}\left\{x \in \Lambda_L : Q^\omega(x) \le q\right\} / L^{d-1}; \tag{8}$$

the ergodicity of Q^ω guarantees the existence and non-randomness of this limit. Then there holds /6/

<u>Proposition 5:</u> Let Q^ω as in Proposition 4 and $c_{p,d}$, $c'_{p,d}$ as in Proposition 3. Then, for $E < 0$

$$n^s(E) \le \int_{-\infty}^{0} \min\left\{c_{p,d}\,(-q)^{d-1}, c'_{p,d}\,(-E)^{-1/2}\,(-q)^d\right\} d\Psi_Q(q). \tag{9}$$

If the function Q^ω is periodic, then the additional spectrum caused by the b.c. is absolutely continuous /4/, and generally the ergodicity of Q^ω implies that the discrete spectrum of H^ω_Q is almost surely empty /9/. Furthermore, in many typical examples of ergodic Q^ω (e.g. potentials of the substitutional alloy type) the surface states form a dense pure point spectrum.

Now we assume $H^\omega = H^\omega_Q + W^\omega$, where the external potential W^ω is the restriction of some ergodic (with respect to $\mathbb{R}^d$ or $\mathbb{Z}^d_a = \mathbb{Z}^{d-1}_a \times (a_d \mathbb{Z})$, $a_d > 0$) function U^ω defined on $\mathbb{R}^d$ onto $\mathbb{R}^d_+$. If U^ω obeys a strengthened ergodicity property, namely that any measurable set of realizations which is invariant under translations $U^\omega(x) \mapsto U^\omega(x_1+y_1,\dots,x_{d-1}+y_{d-1},x_d)$, with y running over $\mathbb{R}^{d-1}$ resp. $\mathbb{Z}^{d-1}_a$, has probability 0 or 1, then the methods outlined in /2/ lead us to pseudodifferential equations which are ergodic with respect to $\mathbb{R}^{d-1}$ resp. $\mathbb{Z}^{d-1}_a$. Arguments of Refs. 7 and 8 suggest that in this case - at least if Q^ω and U^ω are sufficiently well-behaved - one has to expect non-random surface spectrum and density of surface states below the infimum and in the gaps of the spectrum of $-\Delta + U^\omega$: For $E_1 < E_2$ lying in the same gap one will have

$$n^s(E_2) - n^s(E_1) = \lim_{L\to\infty} \big(N(H^\omega_{Q,L} + W^\omega, E_2) - N(H^\omega_{Q,L} + W^\omega, E_1)\big)/L^{d-1}. \tag{10}$$

If the domain G is of the kind $\mathbb{R}^{d-m} \times (\mathbb{R}_+)^m$, with $m < d$, then there can occur new states at the edges of the domain. If the potential and the b.c. satisfy sufficiently strong ergodicity conditions then one can define a series of densities of edge states which will be again non-random (cf. /12/). This reflects the fact that the electronic properties of real ergodic systems - such as (macroscopically) homogeneous crystals, polycrystals, glasses, alloys etc. - are sample-independent.

3. Macroscopic surface potentials

In this section we consider the case of surface potentials which are varying very slowly, so that they can be assumed as being constant in any microscopic region. Such potentials can occur e.g. in metals as an effect of an external electrostatic field, or in containers with walls consisting of several materials. We start with the simplest possibility: Let the region G be the column $\Lambda_L \times \mathbb{R}_+$, with Λ_L as in the previous section, and consider the operator $H_{Q,L} = -\Delta$ inside $\Lambda_L \times \mathbb{R}_+$, with b.c.

$$\frac{\partial u}{\partial x_d}(x,0) = Q(x/L)\, u(x,0), \quad x \in \Lambda_L, \tag{11}$$

at the basis and Neumann b.c. at $\partial\Lambda_L \times \mathbb{R}_+$. We note that in this case there holds an analog of Proposition 1 (cf. /5/). The density of surface states can be evaluated explicitely:

<u>Proposition 6:</u> Let the function Q be Riemann integrable over Λ_1. Then there holds the asymptotic formula (for $E < 0$)

$$\lim_{L\to\infty} N(H_{Q,L},E)/L^{d-1} =$$

$$\tau_{d-1}\,(2\pi)^{-d+1}\int_{\Lambda_1}((Q_-^2(x) + E)_+)^{(d-1)/2}\, d^{d-1}x, \tag{12}$$

where τ_{d-1} denotes the volume of the unit sphere in $\mathbb{R}^{d-1}$.

Proof: We divide the cube Λ_L into cubes $\Lambda_L^{(i)}$ of length $L/[L^{1/2}]$, where the square brackets denote the integer part. On the columns $\Lambda_L^{(i)} \times \mathbb{R}_+$ we define operators $H^N_{Q,L,i}$ and $H^D_{Q,L,i}$ as $-\Delta$ inside the columns, with b.c. (11) at the basis and Neumann resp. Dirichlet b.c. at $\partial\Lambda_L^{(i)} \times \mathbb{R}_+$ (but only Neumann b.c. at $(\partial\Lambda_L \cap \partial\Lambda_L^{(i)}) \times \mathbb{R}_+$). From Dirichlet-Neumann bracketting arguments (cf. /10/) it follows that, for $E < 0$,

$$\sum_i N(H^D_{Q,L,i}, E) \leq N(H_{Q,L}, E) \leq \sum_i N(H^N_{Q,L,i}, E). \tag{13}$$

Further, we introduce the piecewise constant functions

$$Q_1(x) = \inf\{Q(y/L):\ y \in \Lambda_L^{(i)}\} \quad \text{for } x \in \Lambda_L^{(i)}$$

and

$$Q_2(x) = \sup\{Q(y/L):\ y \in \Lambda_L^{(i)}\} \quad \text{for } x \in \Lambda_L^{(i)},$$

and the operators $H^+_{Q,L,i} = H^N_{Q_1,L,i}$ and $H^-_{Q,L,i} = H^D_{Q_2,L,i}$. Since the eigenvalues of $H^N_{Q,L,i}$ and $H^D_{Q,L,i}$ depend on Q monotonously /4/, we obtain

$$\sum_i N(H^-_{Q,L,i}, E) \leq N(H_{Q,L}, E) \leq \sum_i N(H^+_{Q,L,i}, E). \tag{14}$$

It is easy to see that asymptotically

$$N(H^{\pm}_{Q,L,i}, E) \approx \tau_{d-1}\,(2\pi)^{-d+1}\,(L(Q^2_{1,2}(x) + E))^{(d-1)/2}, \tag{15}$$

(with $x \in \Lambda_L^{(i)}$) if $Q_{1,2}(x) < -(-E)^{1/2}$, and 0 elsewise. Obviously, the sums on the left and the right hand side of (14), divided by L^{d-1}, are lower resp. upper Darboux sums for the integral in (12). Since Q has been assumed to be Riemann integrable, the proof is done.

If we now consider the region $\Lambda_L \times [0, aL]$, with some $a > 0$, and at this region the operator $-\Delta$ with b.c. (11) at the basis and Neumann b.c. at the other surfaces, then the asymptotic formula

(15), and hence (12), remains true, independently of a. Therefore we can conclude that, if G_1 is some simply connected region with smooth boundary in $\mathbb{R}^d$, $G_L = LG_1$, $H_{Q,L} = -\Delta$ inside G_L, with b.c. $\frac{\partial u}{\partial n}(x) = Q(x/L)u(x)$, where $\frac{\partial}{\partial n}$ denotes the inner normal derivative and Q is a Riemann integrable function on ∂G_1, then there holds the following asymptotic formula

$$\lim_{L\to\infty} N(H_{Q,L},E)/L^{d-1} =$$

$$\tau_{d-1}\,(2\pi)^{-d+1}\int_{\partial G_1}((Q_-^2(x)+E)_+)^{(d-1)/2}\,dS, \tag{16}$$

with dS denoting the surface element of ∂G_1. This can be proven by considering only a thin layer $\tilde{G}_1$ near ∂G_1 instead of the whole region G_1, and dividing the blown up region $L\tilde{G}_1$ into columns of width $\sim L^{1/2}$ perpendicular to ∂G_L, where formula (15) holds.

References

1. H. Englisch and P. Šeba, The stability of the Dirichlet and Neumann boundary, Rept. Math. Phys. 23(1986) 73.
2. L. Hörmander, Pseudo-differential operators and non-elliptic boundary problems, Ann. Math. 83(1966) 129.
3. M. Schröder, On the spectrum of Schrödinger operators at the half-space with a certain class of boundary conditions, to appear in Z. Anal. Anw.
4. H. Englisch, M. Schröder and P. Šeba, The free Laplacian with attractive boundary conditions, to appear in Ann. Inst. H. Poincaré.
5. M. Schröder, Spektraleigenschaften von Schrödinger-Operatoren mit zufälligen ergodischen Randbedingungen, in: Proc. 2nd Conf. Stochastic Analysis (IH Zwickau 1986).
6. M. Schröder, Estimates on the spectrum of Schrödinger operators with attractive boundary conditions, submitted to Math. Nachr.
7. L.A. Pastur and A.L. Figotin, Ergodic properties of the distribution of the eigenvalues of certain classes of random selfadjoint operators, Sel. Math. Sov. 3(1984) 69.
8. W. Kirsch and F. Martinelli, On the density of states of Schrödinger operators with a random potential,J.Phys.A15(1982) 2139.

9. H. Englisch and M. Schröder, Schrödinger operators with random boundary conditions, to appear in: Proc. Int. Sem. Localization in Disordered Systems (Teubner, Leipzig 1987).
10. M. Reed and B. Simon, Methods of Modern Mathematical Physics IV (Academic Press, New York - San Francisco - London 1978).
11. B. Simon, Functional Integration and Quantum Physics (Academic Press, New York - San Francisco - London 1979).
12. H. Englisch and M. Schröder, Bose condensation in disordered systems. II. Surface and bound states, submitted to Physica A.
13. H. Englisch, Private communication.

QUANTUM JUNCTIONS AND THE SELF-ADJOINT EXTENSIONS THEORY

P.Exner*, P. Šeba**

Joint Institute for Nuclear Research; Laboratory of Theoretical Physics, 141 980 Dubna, USSR

The aim of the presented paper is to describe a simple model of low-energy scattering on a quantum point contact.

Let us first outline the physical background of the model. For a metallic contact,the common wishdom suggests a linear relation between the applied voltage and the current (the Ohm's law). This remains true, however, only if the size of the contact is large enough. Once its diameter becomes comparable with the mean free path of the electrons in the metal scattering effects appear which add a nonlinear contribution to the current.

Effects of this type were measured first at Kharkhov in a pioneering experiment by Igor Yanson [1], which gave rise to a new research branch called now "Point Contact Spectroscopy". The measured nonlinearity represents usually a few promile to a few percent of the total current and is visible in the differential resistance dU/dI. The second derivative d^2U/dI^2 exhibits typically a more complicated shape with peaks corresponding to the electron-phonon interaction in the metal involved.

There are two basic types of the point contact experiments. In the first of them, dubbed a pressure-type contact, a sharply tipped wire is adjusted by a screw against a flat metallic

---

* On leave of absence from the Nuclear Physics Institute, Czechoslovac Academy of Sciences, 25068 Řež near Prague, Czechoslovakia.

** On leave of absence from Nuclear Centre,Charles University, V Holešovičkách 2, Prague 8, Czechoslovakia

surface. The second type consists of two thin metallic films separated by an insulating (oxide) layer which is perforated at one point. In both two types the contact region is typically a few micrometers in diameter.

The "microscopic" theory of these contacts is a very complicated matter and we are not going to discuss it here. We shall restrict our attention to the long-wave-limit situation only, when the de Broglie wavelength of the electrons is muchlonger than the contact diameter. In this case it is reasonable to expect that the contact scattering would not depend on the detailed shape of the contact region, but rather on the global geometry of the experiment. Our aim here is to show that the "free" quantum mechanics together with the geometry of the contact are alone responsible for the measured shape of the current-voltage characteristics in the long-wave limit.

1. A model for the thin-film contact

We consider here the simplest possible quantum-mechanical model [2] in which a free electron moves on a manifold consisting of two planes connected at one point (Fig.1). The state Hilbert space $\mathcal{H}$ of our problem is a sum of the spaces corresponding to the upper and lower plane respectively: $\mathcal{H} = L^2(\mathbb{R}^2)\oplus L^2(\mathbb{R}^2)$. To construct the quantum Hamiltonian we start with the operator

$$H = H_{o,1} \oplus H_{o,2} \qquad (1.1)$$

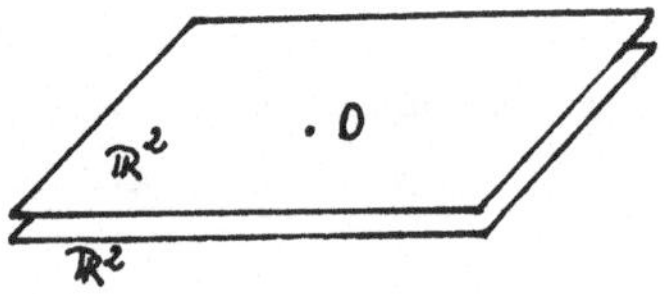

Fig.1

where $H_{o,j} = -\Delta$ with $D(H_{o,j}) = C_o^{\infty}(\mathbb{R}^2 \setminus \{0\})$, $j = 1,2$, assuming the connection point to be placed at the origin on each of the planes ; the particle is supposed to be free outside the connection point. The deficiency indices of the operators $H_{o,j}$ are known to be (1,1) [3], hence the deficiency indices of H_o are (2,2) and it has a four parameter family of self-adjoint extensions. To construct these extensions we proceed in a standard way. The deficiency subspaces $\mathcal{K}^{\pm} = \mathrm{Ker}(H_0^{*} \pm i\mathbb{I})$ are spanned by the vectors $\varphi_k^{\pm}$, $k = 1,2$, where

$$\varphi_k^{+} = (f_0, 0) \;, \qquad \varphi_k^{-} = (0, f_0) \tag{1.2}$$

with

$$f_0(x) = H_0^{(1)}(\varepsilon x) \;, \quad \varepsilon = e^{i\pi/4} \tag{1.3}$$

and φ_k^{-} are complex conjugated to φ_k^{+}. Introducing the polar coordinates in each of the two planes and decomposing the Hilbert space

$$L^2(\mathbb{R}^2) = \bigoplus_{m=-\infty}^{\infty} L^2(\mathbb{R}^+, r\,dr) \otimes \{Y_m\}_{\mathrm{lin}} \tag{1.4}$$

where

$$Y_m(\varphi) = (2\pi)^{1/2} e^{im\varphi} \tag{1.5}$$

we can decompose the operators $H_{0,j}$ as

$$H_{0,j} = \bigoplus_{m=-\infty}^{\infty} h_{m,j} \otimes \mathbb{I} \tag{1.6}$$

where

$$h_{m,j} = -\frac{d^2}{dr^2} - \frac{1}{r}\frac{d}{dr} + \frac{m^2}{r^2} \;; \quad D(h_{m,j}) = C_0^{\infty}(\mathbb{R} \setminus \{0\}) \,. \tag{1.7}$$

All the operators $h_{m,j}$ are e.s.a. for $m \neq 0$ (see [3], Sec. X.1). We get therefore

Proposition 1:

All the self-adjoint extensions of the operator H_0 are of the form

$$H_U = K_U \oplus \bar{h} \,, \tag{1.8}$$

where

$$h = \Big[\bigoplus_{m \neq 0} h_{m,1} \otimes \mathbb{I} \Big] \oplus \Big[\bigoplus_{l \neq 0} \mathbb{I} \otimes h_{l,2} \Big] \tag{1.9}$$

and K_U is a self-adjoint extension of the operator

$$K_0 = (h_{0,1} \otimes \mathbb{I}) \oplus (\mathbb{I} \otimes h_{0,2}) \tag{1.10}$$

with the domain

$$D(K_0) = \Big\{ \varphi = (f_1, f_2) : \; f_j(x) = f_j(|x|) \,, \; f_j \in C_0^{\infty}(\mathbb{R}^+ \setminus \{0\}) \Big\}.$$

(Here U denotes a 2x2 unitary matrix U which is usually used in the von Neumann theory to parametrize self-adjoint extensions.)

Hence it is only necessary to find the operators K_U. The most simple way to do it is to use boundary conditions. The deficiency functions are, however, singular at the origin. Therefore the corresponding boundary conditions must be written in terms of regularized boundary values [4]

$$L_0(f) = \lim_{r\to 0} \frac{f(r)}{\ln r} \quad ; \quad L_1(f) = \lim_{r\to 0} [f(r) - L_0(f) \ln r] .$$

Proposition 2

Every self-adjoint extension K_U is uniquely specified by the following boundary conditions: $f = (f_1, f_2) \in D(K_U)$ iff

$$f_i \in \left\{ f \in L^2(\mathbb{R}^+, rdr) ;\ f, f' \in AC(\mathbb{R}^+) \text{ and } f'' + \frac{1}{r} f' \in L^2(\mathbb{R}^+, rdr) \right\}$$

for $i = 1,2$ and

(i)
$$\begin{aligned} L_0(f_1) &= aL_0(f_2) + bL_1(f_2) \\ L_1(f_1) &= cL_0(f_2) + dL_1(f_2) \end{aligned} \qquad (1.11)$$

where the coefficients are given by

$$\begin{aligned} a &= u_{12}^{-1}[\chi(u_{11}-1) + \bar{\chi}(\det(U) - u_{22})] \\ b &= \frac{2i}{\pi} u_{12}^{-1}[1 - \operatorname{tr}(U) + \det(U)] \\ c &= \frac{\pi i}{2} u_{12}^{-1}[\chi^2 + \chi\bar{\chi} \operatorname{tr}(U) + (\bar{\chi})^2 \det(U)] \\ d &= u_{12}^{-1}[\chi(1-u_{22}) + \bar{\chi}(u_{11} - \det(U))] \end{aligned} \qquad (1.12)$$

and u_{ij} are matrix elements of a nondiagonal unitary 2x2 matrix, or

(ii)
$$\begin{aligned} L_0(f_1) &= AL_1(f_1) \\ L_0(f_2) &= BL_1(f_2) \end{aligned} \qquad (1.13)$$

with $A, B \in \mathbb{R}$. Here $\chi = L_1(f_0) = \frac{1}{2} + \frac{2i}{\pi}(\gamma - \ln 2)$ and $\gamma = 0.577..$ is the Euler's constant.

Proof: See [2].

Remark: The extensions defined by the boundary conditions (ii) are physically not interesting because they lead to Hamiltonians which have are a direct sum and hence describe a system in which the two planes are completely separated. In what follows we restrict ourselves therefore only to extensions given by (i).

We suppose that the two planes are physically equivalent, i.e., we restrict our attention to the Hamiltonians H_U which commute with the modified parity operator P exchanging the planes

$$P : (f_1, f_2) \longrightarrow (f_2, f_1) \quad ; \quad (f_1, f_2) \in \mathcal{H} \; . \tag{1.14}$$

It can be shown [5] that all the Hamiltonians which fulfil

$$PH_U \subset H_U P \tag{1.15}$$

form a two parameter subfamily corresponding to symmetric matrices U. This subfamily can be described by the boundary conditions (i) from the Proposition 2 with the coefficients a,b,c,d given by

$$\begin{aligned} a = -d &= \frac{\cos(\beta) + \cos(\xi) - \sin(\xi)}{\sin(\xi)} \\ b &= 2^{1/2}\,\frac{\cos(\beta) + \cos(\xi)}{\sin(\beta)} \\ c &= 2^{1/2}\,\frac{\sin(\xi) - \cos(\beta)}{\sin(\beta)} \end{aligned} \tag{1.16}$$

where $\beta, \xi \in (0, 2\pi]$ are two real parameters.

Let us now investigate scattering of the particle on the connection point. Our aim is to find the transmission probability from the upper to the lower plane. Using the time-independent approach we start with the function $f = (f_1, f_2)$, where

$$\begin{aligned} f_1(r) &= H_0^{(2)}(kr) + A(k)\, H_0^{(1)}(kr) \\ f_2(r) &= B(k)\, H_0^{(1)}(kr) \end{aligned} \tag{1.17}$$

and demand it to belong locally to $D(H_U)$. A simple calculation yields the coefficients $A(k), B(k)$ which are expressed as

$$A(k) = \frac{c - 2a\ (\gamma + \ln \frac{k}{2}) - b\ [\ \frac{\pi^2}{4} + (\gamma + \ln \frac{k}{2})^2]}{c - 2a\ (\frac{\pi}{2i} + \gamma + \ln \frac{k}{2}) - b\ (\frac{\pi}{2i} + \gamma + \ln \frac{k}{2})^2}$$

$$B(k) = \frac{i\pi}{c - 2a\ (\frac{\pi}{2} + \gamma + \ln \frac{k}{2}) - b\ (\frac{\pi}{2i} + \gamma + \ln \frac{k}{2})^2}$$

A(k) and B(k) are the reflection and transmission coefficients respectively. It can be easily seen that they fulfill

$$|A(k)|^2 + |B(k)|^2 = 1. \tag{1.18}$$

Thus we have obtained a nontrivial particle transmission from the upper to the lower plane.

Before comparing the obtained transmission coefficient with the experimental results we describe in short one more type of a quantum point contact model, which corresponds to the pressure type experiments.

2. A model for the pressure-type contacts

As already mentioned, in the pressure-type experiments a thin wire is adjusted against a flat metallic surface. Fixing the basic geometry and supposing that the linear dimension of the contact is zero we obtain the simplest model in which the electron moves in a manifold consisting of a half-line connected to a plane (Fig.2). The corresponding state Hilbert space is now given as

$$\mathcal{H} = L^2(\mathbb{R}^-) \oplus L^2(\mathbb{R}^2).$$

Such a quantum system has been discussed in [6] and we summarize here the results.

Since the electron is again supposed to be free outside the connection point the starting operator H_0 is given by (1.1) with

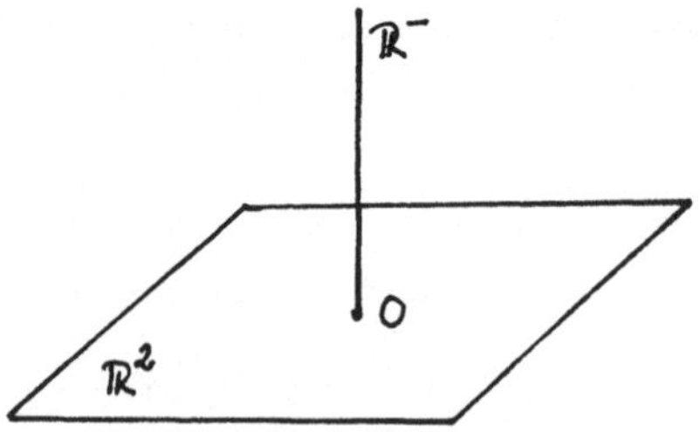

Fig.2

$$H_{0,1} = -\frac{d^2}{dx^2} \quad ; \quad D(H_{0,1}) = C_0^\infty(\mathbb{R}^- \setminus \{0\}) \tag{2.1a}$$

$$H_{0,2} = -\Delta \quad ; \quad D(H_{0,2}) = C_0^\infty(\mathbb{R}^2 \setminus \{0\}) \ . \tag{2.1b}$$

The operator H_0 has deficiency indices (1,1) and the corresponding deficiency subspaces are spanned by vectors

$$\varphi_1^+ = (f_1, 0) \quad ; \quad \varphi_2^+ = (0, f_0) \quad ; \qquad \varphi^- = \overline{\varphi^+}$$

where $f_1(x) = 8^{1/4} \exp(\bar{\varepsilon}x)$, and f_0 is given by (1.3). Using the von Neumann theory we get a four-parameter family of possible self-adjoint extensions of the operator H_0, which is parametrized as usually by 2x2 unitary matrices U. Comparing to the previous case, one cannot eliminate some of the possible Hamiltonians using the symmetry of the manifold. Physically it is, however, reasonable to use only the Hamiltonians leading to a time-invariant particle dynamics. Mathematically speaking this means to use only such extensions H_U of H_0 for which the equivalence

$$f \in D(H_U) \iff \bar{f} \in D(H_U) \tag{2.3}$$

holds. It can be easily shown that this condition leads to a three-parameter family of admissible Hamiltonians corresponding to the matrices U whose elements fulfil $u_{12} = u_{21}$.

These extensions are described by

<u>Proposition 3</u>

Every self-adjoint extension fulfilling (2.3) is specified uniquely by the following boundary conditions.

If $f = (f_1, f_2)$ belongs to $D(H_U)$ then

$$\begin{aligned} L_0(f_2) &= Af_1(0) + Bf_1'(0) \\ L_1(f_2) &= Cf_1(0) + Df_1'(0) \end{aligned} \tag{2.4}$$

where the parameters A,B,C,B are given by

$$A = \frac{8^{1/4}}{\pi} \quad \frac{\sin(\xi + \frac{\pi}{4}) - \cos(\beta)\sin(\alpha + \frac{\pi}{4})}{\sin(\beta)}$$

$$B = \frac{8^{1/4}}{\pi} \quad \frac{\sin(\alpha)\cos(\beta) - \sin(\xi)}{\sin(\beta)}$$

$$C = \frac{[\frac{1}{2} + \frac{2}{\pi}(\gamma - \ln 2)](\cos\varphi + \sin\alpha\cos\beta) + [\frac{1}{2} - \frac{2}{\pi}(\gamma - \ln 2)](\sin\varphi + \cos\alpha\cos\beta)}{8^{1/4}\sin\beta}$$

$$D = -\frac{8^{1/4}}{2} \cdot \frac{\frac{1}{2}(\cos\xi + \cos\alpha\cos\beta) + \frac{2}{\pi}(\gamma - \ln 2)(\sin\xi - \sin\alpha\cos\beta)}{\sin\beta}$$

with $\alpha, \xi \in [0, 2\pi]$ and $\beta \in (0, 2\pi)$.

Let us now investigate the transition from $\mathbb{R}^-$ to $\mathbb{R}^2$ in the framework of time-independent scattering theory Using the boundary conditions (2.4) and requiring the function $f = (\varphi_1, \varphi_2)$ with

$$\begin{aligned} \varphi_1(x) &= e^{ikx} + A(k)\, e^{-ikx} \\ \varphi_2(r) &= B(k)\, H_0^{(1)}(kr) \end{aligned} \tag{2.5}$$

to belong locally to $D(H_U)$, we find

$$A(k) = \frac{\frac{2i}{\Pi}\left[\left(\ln\frac{k}{2} + \gamma\right)\left(A + ikB\right) - C - ikD\right] + A + ikB}{\frac{2i}{\Pi}\left[\left(\ln\frac{k}{2} + \gamma\right)\left(-A + ikB\right) + C - ikD\right] - A + ikB}$$

$$B(k) = \frac{2ik(BC - AD)}{\frac{2i}{\Pi}\left[\left(\ln\frac{k}{2} + \gamma\right)\left(-A + ikB\right) + C - ikD\right] - A + ikB} \tag{2.6}$$

Here $A(k)$ and $B(k)$ are again the reflection and transition coefficients, respectively.

Let us now turn to the experiment and compare the calculated quantum transition coefficients with the measured nonlinearity of the current-voltage characteristic.

3. Comparison with the experiment.

Before comparing our results with the experimental data we have to relate the quantum transition coefficients to the measured resistance of the point contact. This can be done using a formula well known from the electrical transport theory [7]. If the metals involved have the same Fermi energies then the current through the contact is given by

$$I = -\frac{2e}{\hbar}\int_{0}^{\infty} T(E)\,[f_T(E) - f_T(E - eU)]\,dE \qquad (2.7)$$

where e is the electron charge, U is the applied voltage and

$$f_T(E) = \left[1 + \exp\left(\frac{E - E_F}{kT}\right)\right]^{-1} \qquad (2.8)$$

is the metal electron-gas density at the temperature T and Fermi energy E_F. T(E) is the quantum transition probability which is given simply by

$$T(E) = |B(k^2)|^2 \qquad (2.9)$$

where B(k) are the transition coefficients calculated in the Sections 1 and 2, respectively.

The formula (2.7) becomes particularly simple in the zero-temperature limit, when $f_T(E)$ becomes a step function

$$I = -\frac{2e}{\hbar}\int_{E_F}^{E_F+eU} T(E)\,dE \ . \qquad (2.10)$$

Differenciating it by U we get finally the sought formula for the differential resistance

$$\frac{dU}{dI} = \frac{\hbar}{2e^2}\,T(E_F + eU)^{-1} \qquad (2.11)$$

To be just we have to mention at this point that the formula (2.11) has been challenged, however, the alternative proposed in Ref.8 differs only by a constant which is not important for our purpose here.

In the models described in Sections 1 and 2 there are two or three adjustable parameters, respectively, with the help of which we can fit the basic nonlinearity of the current-voltage characteristic measured in the experiment. In order to give an example we plot on Figures 3 and 4 the resistance corresponding via the formula (2.11) to the quantum models constructed in Sections 1 and 2 . The same plots are transformed on Figures 5 and 6 to the logarithmic scale. One has to compare these curves with the voltage dependence of the resistance measured in the point-contact experiment [9] which is plotted on the Figures 7 and 8. One can see that the shape of the resistance obtained in our models fits well the experimental curves. Moreover, the resistance curves obtained in the two-plane and pressure-type models are very similar and are in fact determined only by the geometry of the manifolds involved.

On the other hand the models under consideration cannot describe a more complicated structure observed in the second derivatives of the current-voltage characteristic which is connected with the particular electron-phonon interaction in the metal used. One can, however, add an internal Hilbert space $\mathcal{H}_{in}$ to to the state Hilbert space $\mathcal{H}$ and model in such a way for instance the polaron states localized in the contact region. Constructing the self-adjoint extensions in the larger Hilbert space one obtaines a more complicated structure of the transition coeficients and hence also a more complicated structure of the resistance plot. Adding the internal space it is possible to mathematically combine the global manifold geometry, which is responsible for the dU/dI plot with the detailed electron-phonon interaction being responsible for the complicated structure of the

d^2U/dI^2 plot. The model obtained in such a way is, however, very similar to the zero-range interaction models described in this volume and we omit the details here.

References

1. I.K.Yanson: Zh.Eksp.Teor.Fiz. 66 (1974) 1035
2. P.Exner, P.Šeba: Lett.Math.Phys. 12 (1986) 193
3. M.Reed, B.Simon: Methods of Modern Mathematical Physics II ; Academic Press, New York 1975
4. W.Bulla, F.Gesztesy: J.Math.Phys. 26 (1985) 2520
5. P.Exner, P.Šeba: A simple model of thin-film contacts in two and three dimensions. Czech.J.Phys. B38 (1988) in press
6. P.Exner, P.Šeba: J.Math.Phys.28 (1987) 386
7 K.C.Kao, W.Huang: Electrical Transport in Solids. Pergamon Press, New York, 1981
8. R.Landauer: Phys.Lett.A. 85 (1981) 91
9 A.G.M. Jansen, A.P. van Gelder, P. Wyder: J.Phys.C 13 (1980) 6073

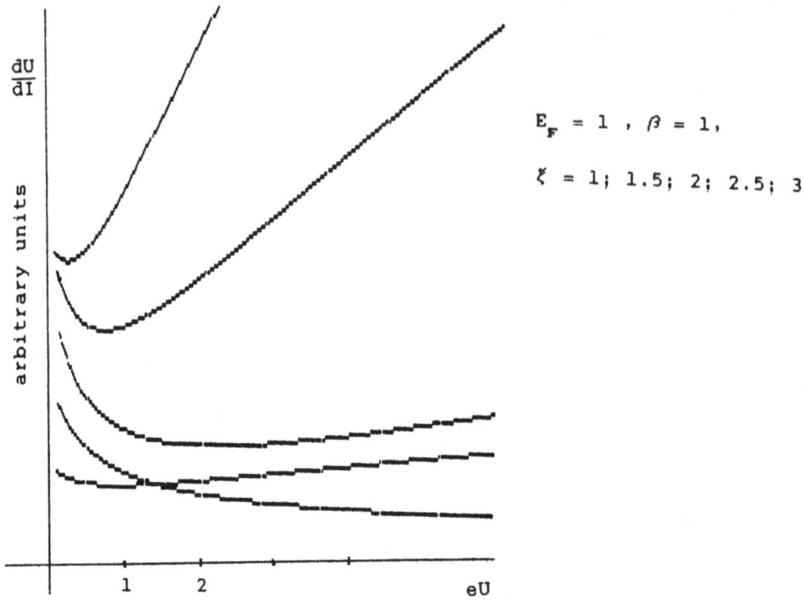

Fig.3: The resistance plots obtained in the thin-film model for various self-adjoint extensions.

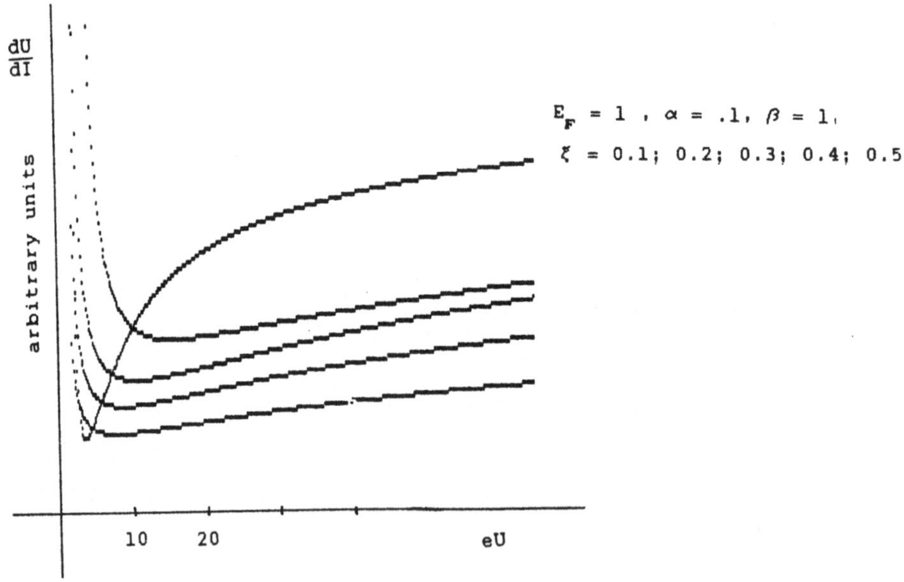

Fig.4: The resistance plots obtained in the pressure-type model

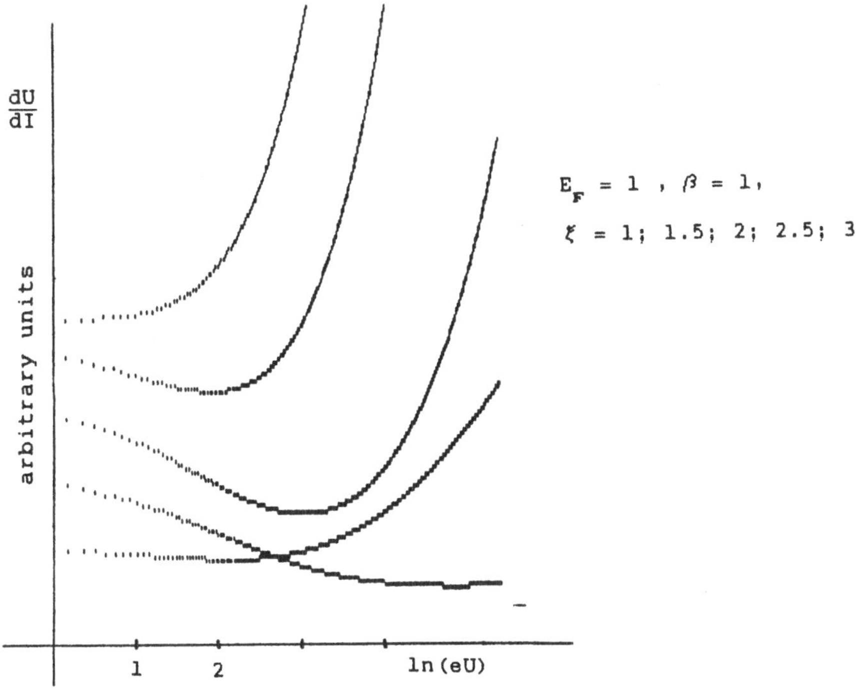

Fig.5: The resistance plots obtained in the thin-film model for various self-adjoint extensions: the logarithmic scale

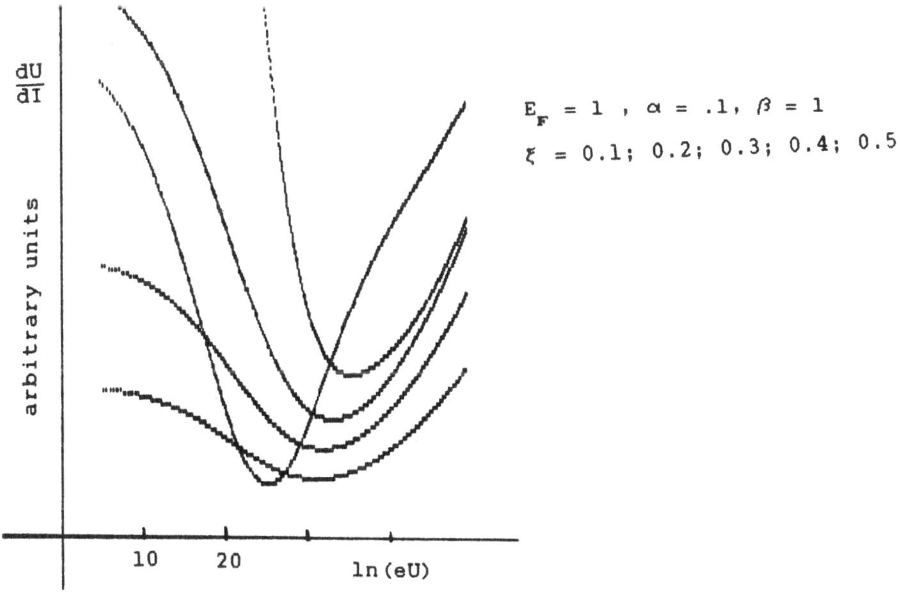

Fig.6: The resistance plots obtained in the pressure-type model: the logarithmic scale

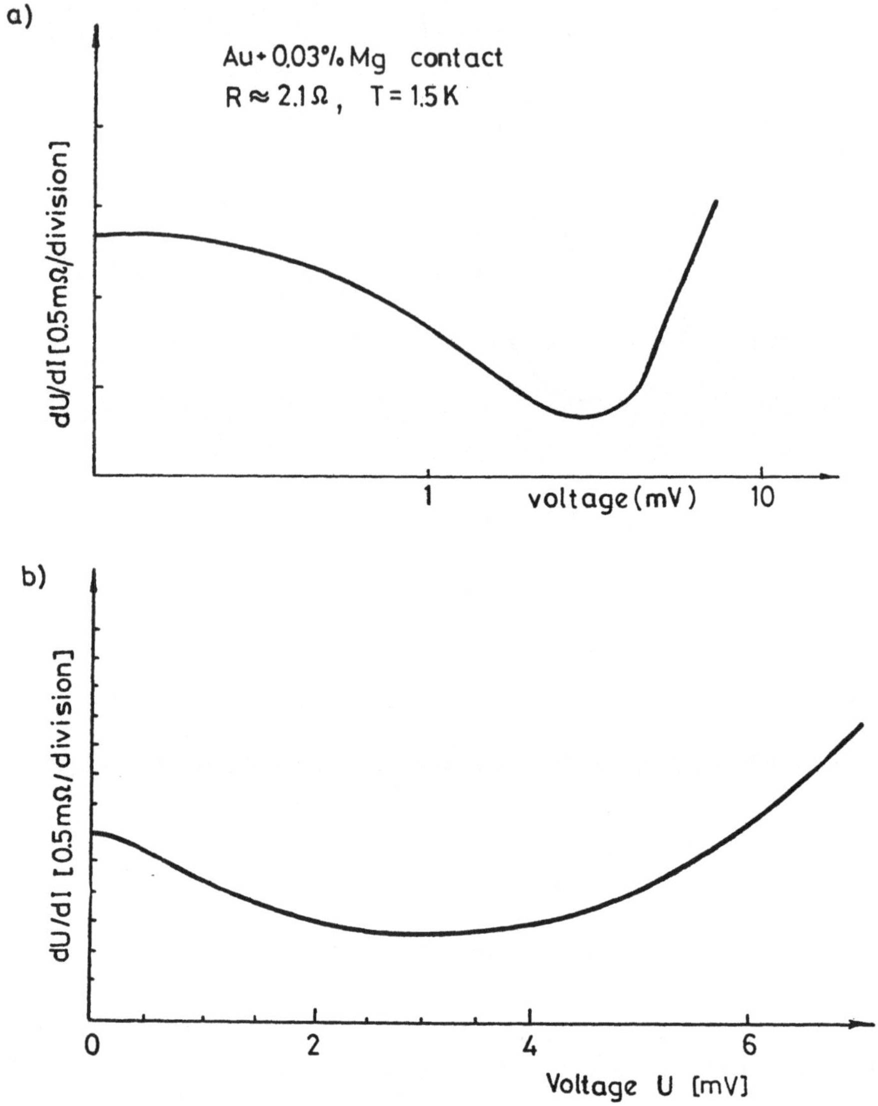

Fig.7: A typical resistance plot obtained in a thin-film point contact
a) the logarithmic scale
b) the linear scale

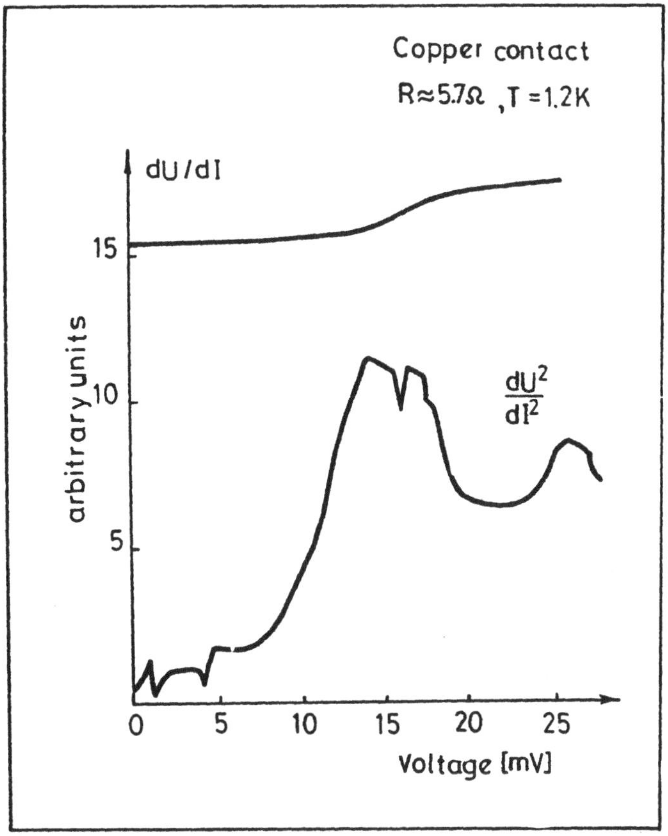

Fig.8: A typical resistance plot obtained in a pressure type point contant experiment

THE EXTENSION THEORY AND DIFFRACTION PROBLEMS

I. Yu. Popov
Department of Mathematical
and Computational Physics, Physical Faculty,
Leningrad State University, 1 Maya 100,
Petrodvoretz, Leningrad 198904

A model describing the scattering problem on a surface with a small opening or a narrow slit is discussed in the framework of the extension theory. This approach simplifies the problem considerably and makes it explicitly solvable in many cases. A comparison to the "real" diffraction problem is also presented.

Most of interesting diffraction problems cannot be solved in an analytic form; let us recall boundary problems for Helmholtz equation in domains with small openings, slit antennas, a circular resonator with a running wave etc. From the physical point of view, however, it is evident that the wave dynamics in such complicated domains is a superposition, in some sense, of the "local" dynamics relative to components of the domain, and that the spectral properties of Laplace operators in such domains depend on the spectral properties of the "local" operators and on the quality of the channel coupling. The aim of the present lecture is to give an exact meaning to these heuristic statements in the framework of the extensions theory.

1. The Neumann problem

Let Ω^{in} be a compact domain in R^3 with a smooth boundary, $\Omega^{ex} = R^3 \setminus \Omega^{in}$. Consider the Laplace operators $-\Delta^{in}$, $-\Delta^{ex}$ in

Ω^{in}, Ω^{ex} with the Neumann boundary condition. Let us restrict these operators to the linear sets of smooth functions vanishing near a point $x_0 \in \partial\Omega$. The restricted operators will be denoted as $-\Delta_0^{in}$, $-\Delta_0^{ex}$.

<u>Lemma</u> [1]: The operator $-\Delta_0^{in}$ has deficiency indices (1, 1). The deficiency element which corresponds to the complex point λ_0 coincides with the Green's function of the internal Neumann problem $G^{in}(x, x_0, \lambda_0)$ which has a pole at x_0.

This lemma can be obtained from the asymptotic behaviour of the Green's function near the pole on $\partial\Omega$:

$$G^{in}(x, x_0, \lambda) = \frac{1}{2\pi|x - x_0|} + \frac{H}{4\pi} \ln \frac{1}{|x - x_0|} +$$

$$+ (\lambda - \lambda_0) \sum_n \frac{|\varphi_n(x_0)|^2}{(\lambda_n - \lambda)(\lambda_n - \lambda_0)} + c^{in}(\lambda_0, x_0) + o(1).$$

Here H is the Gaussian curvature of the boundary at the point x_0, φ_n are the eigenfunctions of $-\Delta^{in}$ corresponding to the eigenvalues λ_n, and λ_0 is a fixed value of the spectral parameter.

An analogous formula holds also for the external domain,

$$G^{ex}(x, x_0, \lambda) = \frac{1}{2\pi|x - x_0|} - \frac{H}{4\pi} \ln \frac{1}{|x - x_0|} +$$

$$+ (\lambda - \lambda_0) \int \frac{|\varphi^{ex}(x_0, \nu)|^2}{(|\nu|^2 - \lambda)(|\nu|^2 - \lambda_0)} d\nu + c^{ex}(\lambda_0, x_0) + o(1)$$

where $\varphi^{ex}(x_0, \nu)$ are the scattered waves in Ω^{ex}. The assertion of the lemma is valid for $-\Delta^{ex}$ too. It is important that the singular parts of these asymptotics do not depend on the spectral parameter.

To construct an extension of $-\Delta_0$, it is necessary to describe the domain of its adjoint operator. For this purpose it is convenient to introduce a "real" basis W^+, W^- in the direct sum of deficiency subspaces in the following way: If h is the normalized deficiency element of $-\Delta_0$ (for the sake of simplicity, we drop temporarily the superscript "in" or "ex"), we set

$$W^+ = (-\Delta - i)^{-1} (-\Delta)\, h, \qquad W^- = (-\Delta - i)^{-1}\, h.$$

In fact, we replace the basis $\{G(x, x_0, i), G(x, x_0, -i)\}$ by the new basis $\{W^- = \operatorname{Im} G(x, x_0, i)(\operatorname{Im} G(x, x_0, i))^{-1}, W^+ = \operatorname{Re} G(x, x_0, i)\}$. The domain of the adjoint operator can be represented in the form:

$$D(-\Delta_0^*) = \{u:\ u = (-\Delta - i)^{-1} v + a_u^+ W^+ + a_u^- W^-,\ v \perp h\}.$$

The boundary-form value for $u, v \in D(-\Delta_0^*)$ equals to

$$J[u, v] = (-\Delta_0^* u, v) - (u, -\Delta_0^* v) = 2\,(a_u^- \overline{a_v^+} - a_u^+ \overline{a_v^-}).$$

One can prove the last equality in the abstract form taking into account the relations $-\Delta_0^* W^+ = W^-$ and $-\Delta_0^* W^- = -W^+$. Alternatively, one can integrate by parts in our concrete case using the asymptotic formulas

$$W^+ = \frac{1}{2\pi|x - x_0|} \pm \frac{H}{4\pi} \ln \frac{1}{|x - x_0|} + O(1),$$

$$W^- = 1 + O(1).$$

For any $\varepsilon \to 0$, we get

$$J[u, v] = \int_{\Sigma_\varepsilon} \left(\frac{\partial u}{\partial n} \bar{v} - u \frac{\overline{\partial v}}{\partial n} \right) ds,$$

where $\Sigma_\varepsilon = \{x: |x - x_0| = \varepsilon,\ x \in \Omega\}$. The limit $\varepsilon \to 0$ yields the above formula.

Consider now the operator $\overline{-(\Delta_0^{in} + \Delta_0^{ex})}$. Its boundary form coincides with the sum of boundary forms of internal and external operators.

The domain of a selfadjoint extension is a linear set of functions, for which the boundary form vanishes. There exist an extensive family of boundary conditions leading to selfadjoint operators, but we restrict further on to one of them. It is determined by the conditions

$$a^{-,in} = a^{-,ex}, \quad -a^{+,in} = a^{+,ex} \qquad (1)$$

Restricting the adjoint operator $-(\Delta_0^{in} + \Delta_0^{ex})^*$ to the domain containing functions which fulfil the boundary condition (1) we get a selfadjoint extension $-\widetilde{\Delta}$ of $-(\Delta_0^{in} + \Delta_0^{ex})$.

We can construct the S-matrix corresponding to this operator. First of all we note that the Green's function has a form:

$$G^{\substack{in\\ex}}(x, x_0, \lambda) = (\lambda - \lambda_0) \int G^{\substack{in\\ex}}(x, s, \lambda)\, G^{\substack{in\\ex}}(s, x_0, \lambda_0)\, ds$$

$$+ i \operatorname{Im} G^{\substack{in\\ex}}(x, x_0, \lambda_0) + \operatorname{Re} G^{\substack{in\\ex}}(x, x_0, \lambda_0),$$

$$\operatorname{Im} \lambda_0 \neq 0.$$

It is now evident that there exist a limit

$$\lim_{x \to x_0} \left(G^{\substack{in\\ex}}(x, x_0, \lambda) - \operatorname{Re} G^{\substack{in\\ex}}(x, x_0, \lambda_0)\right) = D^{\substack{in\\ex}}(\lambda).$$

Here $D^{in,\, ex}(\lambda)$ can be represented as a Schwartz integral, and we get

$$D^{in}(\lambda) = \sum_n \left(\frac{1}{\lambda_n^{in} - \lambda} - \frac{\lambda_n^{in}}{(\lambda_n^{in} - \sigma_0)^2 + \delta_0^2} \right) |\varphi_n^{in}(x_0)|^2,$$

$$\lambda_0 = \sigma_0 + i\,\delta_0,$$

$$D^{ex}(\lambda) = (2\pi)^{-3} \int_0^\infty \sqrt{s}\, ds \, ((s-\lambda)^{-1} -$$

$$- \frac{s}{(s-\sigma_0)^2 + \delta_0^2} \int_{\Sigma_1^{ex}} |\varphi^{ex}(x_0, \nu\sqrt{s})|^2 \, d\nu.$$

Using the usual anzatz for the external part of scattered waves,

$$\psi(x, \nu, \lambda) = \psi^{ex}(x, \nu, \lambda) + f(x_0, \nu, \lambda)\, G^{ex}(x, x_0, \lambda)$$

and using the boundary conditions (1), we obtain

$$\psi(x, \nu, \lambda) = \begin{cases} \dfrac{\psi^{ex}(x_0, \nu, \lambda)}{D^{in}(\lambda) + D^{ex}(\lambda)}\, G^{in}(x, x_0, \lambda), \; x \in \Omega^{in}, \\[2ex] \psi^{ex}(x, \nu, \lambda) \\[1ex] - \dfrac{\psi^{ex}(x_0, \nu, \lambda)}{D^{in}(\lambda) + D^{ex}(\lambda)}\, G^{ex}(x, x_0, \lambda), \; x \in \Omega^{ex}. \end{cases}$$

Hence the S-matrix is given by

$$S(\omega, \nu, k) = S^{ex}(\omega, \nu, k)$$

$$- \frac{i\, k\, \psi^{ex}(x_0, \nu, \lambda)\, \psi^{ex}(x_0, \omega, \lambda)}{2\pi (D^{in}(\lambda) + D^{ex}(\lambda))}, \quad k^2 = \lambda.$$

Here $S^{ex}(\omega, \nu, k)$ is the S-matrix for the unperturbed external problem. This simple formula allows us to find the discrete spectrum and resonances of $-\widetilde{\Delta}$. If the internal and external domains are weakly connected, $|\lambda_0| \to \infty$, the resonances can be approximately calculated (this situation corresponds to the small-opening case-see below). The resonances appear from positive eigenvalues λ_n of the unperturbed internal operator and fulfil

$$k = k_n + \frac{i\,|\varphi_n(x_0)|}{2\lambda_n \|f^{ex}(x_0,\cdot,\lambda)\|_{L_2(\Sigma_1)}(1 + C|\varphi_n(x_0)|\ln|\lambda_0|)}$$

$$+ O(\ln^{-2}|\lambda_0|),$$

where $k_n = \sqrt{\lambda_n}$. It is important that we obtain also an estimate for the imaginary part of the resonance, which characterize the lifetime of this state. Within the standard approach, it is usually a very difficult problem.

It is interesting to compare our model with the real diffraction problem on a surface with a small opening. Provided the extension is fixed, there exists only one free parameter: the point λ_0 for which we have constructed the deficiency elements. It is reasonable to require that the flux through the opening in our model equals to the main part of the real flux as the size of the hole approaches zero. In the two-dimensional case the comparison has been made for many particular diffraction problems (for example, two halfplanes connected through the interval, a half-elliptic resonator connected to a halfplane etc.). It appears that the above conjecture is true [2] if we choose $\lambda_0 = k_0^2$ with $k_0 = 2i\, d^{-1} e^{-\gamma}$, where d is the size of the hole and γ is the Euler's constant. Later this conjecture has been proven by Gottlieb [3].

2. The Dirichlet problem

It is interesting to construct such a model also for the Dirichlet boundary condition. Unfortunately, the approach described above leads to a trivial model. The reason is that the restriction of the Laplace operator to the set of functions vanishing in a neighbourhood of x_0 leads to an essentially selfadjoint operator. Nevertheless, we can build a non-trivial model, adding to the usual Hilbert space L_2 solutions of the Helmholtz equation. These solutions (for instance, a dipole or a higher multipole solution) can become the deficiency elements of the restricted operator. This approach leads, however, to an indefinite-metric space [4].

There exist another way connected with the mixed boundary-condition problem

$$\left(\frac{\partial u}{\partial n} - \sigma u \right)\Big|_{\partial\Omega} = 0.$$

The asymptotics of the corresponding Green's function near the pole x_0 at the boundary is similar to the previous one [5]:

$$G^{\substack{in \\ ex}}(x, x_0, \lambda) = (2\pi |x - x_0|)^{-1} \pm$$

$$\pm (4\pi)^{-1} (H + (2\pi)^{-1}\sigma) \ln (|x - x_0|)^{-1}$$

$$+ O (1) .$$

This property allows us to proceed with the scheme described above. Then we require σ to go to infinity obtaining in this way some results about the perturbed Dirichlet problem. It is crucial for the limit procedure that the Green's function has the following properties:

<u>Theorem</u>:

$$\lim_{\sigma\to\infty} G_\sigma(x, y, k) = G_D (x, y, k); \; x, y \notin \partial\Omega,$$

$$\lim_{\sigma\to\infty} \sigma G_\sigma(x, y, k) = \frac{\partial G_D}{\partial n_x} (x, y, k); \; x \in \partial\Omega , \; y \notin \partial\Omega ,$$

$$\lim_{\substack{y \to x \\ \sigma \to \infty}} \sigma (G_\sigma(x, y, k) - G_\sigma(x, y, k_0)) = 0, \quad x \in \partial\Omega,$$

$$\lim_{\substack{y \to x \\ \sigma \to \infty}} \sigma^2 (G_\sigma(x, y, k) - G_\sigma(x, y, k_0)) =$$

$$= \lim_{y \to x} \left(\frac{\partial^2 G_D}{\partial n_x^2} (x, y, k) - \frac{\partial^2 G_D}{\partial n_x^2} (x, y, k_0)\right) = D(\lambda), \ x \in \partial\Omega.$$

Here $G_D (x, y, k)$ is the Green's function of the Dirichlet problem. All quantities appearing in the theorem exist if the boundary is smooth.

The most natural formula is obtained if we choose the extension specified by the boundary condition

$$a^{-,in} = - a^{-,ex}, \quad a^{+,in} = a^{+,ex}$$

which differs from (1). If $\sigma \to \infty$, we get a formal expression for the scattered waves:

$$\psi(x, \nu, k) = \begin{cases} = \dfrac{\dfrac{\partial \psi^{ex}}{\partial n}(x_0, \nu, k)}{D^{in}(\ ) + D^{ex}(\lambda)} \dfrac{\partial G_D^{in}}{\partial n}(x, x_0, k); \ x \in \Omega^{in}, \\[2ex] \psi^{ex}(x, \nu, k) - \\[2ex] - \dfrac{\dfrac{\partial \psi^{ex}}{\partial n}(x_0, \nu, k)}{D^{in}(\lambda) + D^{ex}(\lambda)} \dfrac{\partial G_D^{in}}{\partial n}(x, x_0, k); \ x \in \Omega^{ex}, \end{cases}$$

Here the function $D^{in}(\lambda)$, $D^{ex}(\lambda)$ depend on the extension parameter λ_0.

We can compare the result with the solution of a real diffraction problem. Let $\Gamma_d = \{x \in \partial\Omega, |x - x_0| < d\}$ be the opening. The scattered wave in this case can be represented as [3]

$$\psi_d(x, \nu, k) = \begin{cases} \int_{\Gamma_d} \psi_d(z, \nu, k) \dfrac{\partial G_D^{in}}{\partial n_\Gamma}(x, z, k)\, dz, & x \in \Omega^{in}, \\ \psi^{ex}(x, \nu, k) + \\ + \int_{\Gamma_d} \psi_d(z, \nu, k) \dfrac{\partial G_D^{ex}}{\partial n}(x, z, k)\, dz, & x \in \Omega^{ex}. \end{cases} \quad (2)$$

Using (2), we calculate first the function $\psi_d(x, \nu, k)$ on the sphere $\Sigma_d = \{x: |x - x_0| = d\}$. After that we employ the Green's function G^C for a ball to compute the value of $\psi_d(x, \nu, k)$ on Γ_d. As a result, we get the following integral equation for ψ_d:

$$\psi_d(x, \nu, k) - \int_{\Gamma_d} \psi_d(z, \nu, k)\, dz \int_{C^{in}} \frac{\partial G_D^{in}}{\partial n}(s, z, k) \frac{\partial G_k^C}{\partial n_C}(s, x)\, ds -$$

$$- \int_{\Gamma_d} \psi_d(z, \nu, k)\, dz \int_{C^{ex}} \frac{\partial G_D^{ex}}{\partial n}(s, z, k) \frac{\partial G_k^C}{\partial n_C}(s, x)\, ds =$$

$$= \int_{C^{ex}} \psi^{ex}(s, \nu, k) \frac{\partial G_k^C}{\partial n_C}(s, x)\, ds,$$

where $C^{in,\, ex} = \Sigma_d \cap \Omega^{in,\, ex}$. For a sufficiently small d, the integral operator appears to be contractive, and we can solve this equation by iterations. The parameter k_0 is chosen in such a way that the flux through the hole in the real and model problem coincide The condition have the form of an algebraic equation:

$$g^{in} + g^{ex} = \left(\int_{C^{ex}} (s - x_0, n_\Gamma) \frac{\partial G_0^C}{\partial n_C} (s, x_0)\, ds \right)^{-1} \left(S_d^{-1} - \int_{C^{in}} \frac{\partial G_0^{in}}{\partial n_\Gamma} (s, x_0) \frac{\partial G_0^C}{\partial n_C} (s, x_0)\, ds - \int_{C^{ex}} \frac{\partial G_0^{ex}}{\partial n_\Gamma} (s, x_0) \frac{\partial G_0^C}{\partial n_C} (s, x_0)\, ds \right),$$

where S_d is the opening area, $G_0^{C,\ in,\ ex}(x, y)$ is the Green's function of the Laplace equation and

$$g^{in,\ ex} = \left(\frac{\partial^2 G_0^{in,\ ex}}{\partial n^2}(x, x_0) - \frac{\partial^2 G_D^{in,\ ex}}{\partial n^2} (x, x_0, k_0)\right)\Big|_{x = x_0}$$

It is essential that the spectral and geometrical characteristics are separated in this equation.

Our model can be generalized to describe the case of a domain with a narrow slit [6]. In this situation it is necessary to impose the boundary conditions on a line. The deficiency indices of the Laplace operator restricted to the set of functions vanishing near the slit are infinite,and the deficiency elements are the potentials of simple layers with the density from the Sobolev's space H_{-1}. In order to satisfy the boundary conditions, which define the sought selfadjoint operator, we have to solve an integral equation. In the cases of simple domains this equation can be solved explicitly. This is true, for example, when the variables in the internal and the external problems separate, and the "connection" line coincides with one of the coordinate axis (a plane with a straight slit, a cylinder or sphere with straight or circular slits [7] etc).

3. A model of coupled resonators

Our construction is particularly useful if one has to investigate a system of coupled resonators [2, 8, 9]. Let us consider the problem of a circular resonator with running wave which can be reduced to the spectral problem for Laplace operator with Neumann boundary condition on a symmetric domain consisting of a several identical circles coupled to a ring (Figure 1). Let λ_n^1 be the eigenvalues of the Laplacian inside the circles, φ_n being the corresponding eigenfunctions. Let further λ_n^0 be the eigenvalues of the Laplacian inside the ring and ψ_n being the corresponding eigenfunctions. Then the dispersion equation for λ has the following form:

Figure 1. The resonator

$$(\lambda-\lambda_0)\sum_n |\varphi_n|^2 (\lambda_n^1-\lambda)^{-1}(\lambda_n^1-\lambda_0)^{-1} +$$

$$+ N(\lambda-\lambda_0)\sum_n |\psi_{Nm+p}|^2 (\lambda^0_{Nm+p}-\lambda)^{-1}(\lambda^0_{Nm+p}-\lambda_0)^{-1} +$$

$$+\sum_m [(N-1)|\psi_{Nm+p}|^2 (\lambda^0_{Nm+p}-\lambda_0)^{-1} -$$

$$-\sum_{j=1}^{N-1} |\psi_{Nm+p+j}|^2 (\lambda^0_{Nm+p+j}-\lambda_0)^{-1} = 0.$$

Here N is the number of circles, p = 0, 1, ... N-1. The roots of this equation are approximately equal to the eigenvalues of corresponding "real" operator for such complicated domain.

References

1. B. S. Pavlov, M. D. Faddeev, Zapisky of Leningrad Branch of Steklov Inst. (LOMI) 126 (1983), 159.
2. M. M. Zimnev, I. Yu. Popov, Soviet J. Vych. Mat. and Mat. Fiz. 27 (1987), 466.
3. V. Yu. Gottlieb, DAN SSSR 287 (1986), 1109.
4. I. Yu. Popov, DAN SSSR 294 (1987), 330.
5. M. Yu. Drozdov, I. Yu. Popov, Leningrad Univ. Vestnik, Ser. 4, No 3 (1987), 93.
6. B. S. Pavlov, I. Yu. Popov, Leningrad Univ. Vestnik No 19 (1983), 36.
7. I. Yu. Popov, Leningrad Univ. Vestnik, No 16 (1984), 79.
8. B. S. Pavlov, I. Yu. Popov, Leningrad Univ. Vestnik, No 4 (1985), 99.
9. B. S. Pavlov, I. Yu. Popov, Leningrad Univ. Vestnik, Ser. 1, No 4 (1986), 105.

HAMILTONIANS WITH ADDITIONAL KINETIC ENERGY TERMS ON HYPERSURFACES

W. Karwowski

Institute of Theoretical Physics, University of Wrocław
Wrocław, Poland

In many cases informations about a physical system can be derived from study of a particle in a field of external potential forces.

Such systems are described by Hamiltonians of the form

$$H = -\Delta + V.$$

Very often this formula has only intuitive value and requires additional input to acquire precise meaning of a self-adjoint operator in Hilbert space.

Sometimes the forces in question are very small away of a set S of measure zero. Then it seems reasonable to consider the idealization. Namely a potential being infinite on S and zero away of S.

The corresponding Hamiltonian is expected to be a limit

as $\varepsilon \to 0$ of $H_\varepsilon = -\Delta + V_\varepsilon$, where V_ε is the multiplication by a bounded measurable function. As $\varepsilon \to 0$ the supports of V_ε shrink to S and the maximum of V_ε goes to infinity.

It seems natural to take for V_ε a sequence converging to the Dirac δ distribution and in fact one uses the symbolic notation of the kind $-\Delta + \int_S \delta(x-y)V(y)dy$ for the limit of H_ε as $\varepsilon \to 0$. It turns out however that unless S has co-dimension 1 one needs a slower growth of V_ε.

In fact it has been shown [1] that with appropriate choice of the sequence V_ε , the H_ε converges in the strong resolvent sense to a positive selfadjoint operator. The problems of this kind include the so called point interactions. They have been extensively studied together with the scattering theory in the forthcoming book on Soluble Models in Quantum Mechanics by S. Albeverio. R. Hoegh-Krohn, H. Holden, F. Gesztezy. The authors consider also Hamiltonians with both the point interaction and that described by the bona fide potentials.

It is characteristic that the "potential supported by the set of Lebesgue measure zero" corresponds to a certain boundary conditions on S.

It is this fact that bears similarity between those "potentials" and the problem we are going to discuss.

Namely we shall also be interested in the Hamiltonians with boundary conditions on the sets of Lebesgue measure zero, but these boundary conditions have different character than those obtained from the potentials.

As we shall see it is possible to give a meaning of a positive selfadjoint operator in $L^2(\mathbb{R}^n, dx)$ to the formal expression

$$- \Delta - \alpha\delta(x_n)\Delta_{n-1} , \tag{1}$$

where Δ is usual Laplace operator and $\Delta_{n-1} = \frac{\partial^2}{\partial x_1^2} + \ldots + \frac{\partial^2}{\partial x_{n-1}^2}$

We gave this formula to ilustrate the origine of the difference between the "potentials" and our operators. It also explains the title of our talk.

Our study will be based on the quadratic forms technics but in the case like that of (1) Gorzelańczyk (private communication) has shown that there is a (strong resolvent limit) of the operators $-\Delta - u(\varepsilon)\chi_{[-\varepsilon,\varepsilon]}(x_n)\Delta_{n-1}$ as $\varepsilon \to 0$, where $\chi_{[-\varepsilon,\varepsilon]}$ is the characteristic function of the interval $[-\varepsilon,\varepsilon]$ and $u(\varepsilon)$ the real function such that $u(\varepsilon) \to \infty$ as $\varepsilon \to 0$.

The motivation for our study is twofold. First it should have applications in acoustic. Second: the surface physics is very complicated. We would like to test our models against the surface physics fenomena. Perhaps there are efects that will fit in the additional kinetic energy scheme rather then, or together with, the potentials. It may prove true or not, but there might be another argument that makes the whole concept more acceptable for the physicists. Namely Dabrowski (priv.com.) claims that the Hamiltonians with the additional kinetic energy on manifolds can be described by the nonlocal potentials.

After this introduction we shall state the main theorem which shows how to define our operators. Then we consider several examples. They are very simple but nevertheless they exhibit effects of the kinetic energy terms.

Let $n > 2$ and Ω_n a nonempty open subset of $\mathbb{R}^n$. Select n constants $C_1,\dots,C_n$ such that for s with $0 < s < n-1$

$$\Omega_s = \{(x_1,\dots,x_n)\in\Omega_n \;;\; x_{s+1} = C_{s+1},\dots,x_n = C_n\} \neq \emptyset .$$

If $f \in L^2(\mathbb{R}^n;dx)$ we put $f^s(x_1,\dots,x_s) \equiv f(x_1,\dots,x_s, C_{s+1},\dots,C_n)$. Also $(x_1,\dots,x_s)\in\Omega_s$ will be the shorthand for

$$(x_1,\dots,x_s,C_{s+1},\dots,C_n)\in\Omega_s .$$

<u>Theorem</u>. The following forms defined on $C_0^1(\Omega_n)$ are positive and closable in $L^2(\Omega_n,dx)$:

$$\varepsilon_k(f,g) = \int_{\Omega_n} \nabla_n f\cdot\nabla_n g\,dx_1 \,\dots\, dx_n \;+$$

$$+\;\alpha_{n-1}\int_{\Omega_{n-1}} \nabla_{n-1}f^{n-1}\cdot\nabla_{n-1}g^{n-1}dx_1\dots dx_{n-1} \;+$$

$$+ \beta_{n-1}\int_{\Omega_{n-1}} f^{n-1}g^{n-1}dx_1\ldots dx_{n-1} + \ldots + \alpha_k\int_{\Omega_k} \nabla_k f^k\cdot\nabla g^k dx_1\ldots dx_k$$

$$+ \beta_k \int_{\Omega_k} f^k g^k dx_1\ldots dx_k + \beta_{k-1}\int_{\Omega_{k-1}} f^{k-1}g^{k-1}dx_1\ldots dx_{k-1},$$

where $k=1,2,\ldots,n;\quad \alpha_1,\ldots,\alpha_{n-1}>0,\ \beta_0,\beta_1,\ldots,\beta_{n-1}>0$

and $\int_\Omega f^0g^0 = f^0g^0 = f(c_1,\ldots,c_n)g(c_1,\ldots,c_n)$.

There is one to one correspondence between positive closed quadratic forms and positive self-adjoint operators. The formal expressions corresponding to the operators given by the above forms read:

$$H_k = -\Delta - \delta(x_n - c_n)(\alpha_{n-1}\Delta_{n-1} - \beta_{n-1}) - \ldots -$$

$$- \delta(x_n-c_n)\ldots\delta(x_{k+1}-c_{k+1})(\alpha_k\Delta_k-\beta_k)+\delta(x_n-c_n)\ldots\delta(x_k-c_k)\beta_{k-1}.$$

Remarks

1) This theorem is a simple corollary to the result of [2]. The proof is based on a version of the imbedding theorem for the Sobolev spaces which implies the inequality:

$$\int_{\Omega_s}\nabla_s f^s\cdot\nabla_s f^s dx_1\ldots dx_s \geq C\int_{\Omega_{s-1}} |f^{s-1}|^2 dx_1\ldots dx_{s-1} = C\|f^{s-1}\|^2_{\Omega_{s-1}}.$$

2) If we put $\alpha_i=0$ for an $i>k$ then the form is not closable.

3) If $k=1$ and $\beta_0>0$ then we get the point interaction for any n.

4) Since $-\Delta < H_k$ we can define $H_{kV} = H_k + V$ where V is any potential small with respect to $-\Delta$ in the Kato sense.

Examples

Put

$$\mathbb{R}^3 = \Omega_3 \ , \quad \Omega_2 = \mathbb{R}^2$$

Let us consider following simple models

1) $H = -\Delta + V(x)$

2) $H = -\Delta + \delta(x_3)\beta$

3) $H = -\Delta - \delta(x_3)(\alpha\Delta_2 - \beta)$

4) $H = -\Delta - \delta(x_3)(\alpha\Delta_2 - \beta) + V(x)$

where $\alpha > 0$, $\beta > 0$, $V(x) = \begin{cases} 0; & x_3 < 0 \\ V; & x_3 > 0 \end{cases}$, $V < 0$.

The resulting boundary conditions are

i) $f(\underline{x},+0) = f(\underline{x},-0)$

ii) $f'(\underline{x},+0) - f'(\underline{x},-0) = (-\alpha\Delta_2 + \beta)f(\underline{x},0)$.

We shall examine propagation of the plane waves. Thus we put

$$f(x) = \begin{cases} e^{ipx} + Ae^{i(\underline{px} - p_3x_3)}, & x_3 < 0 \\ Be^{iqx}, & x_3 > 0, \end{cases}$$

where $\underline{p} = (p_1, p_2)$, $\underline{q} = \underline{p}$, $q_3 = \sqrt{p_3^2 - V}$.

We get

$$A = \frac{(p_3 - q_3) - i(\alpha\underline{p}^2 + \beta)}{(p_3 + q_3) + i(\alpha\underline{p}^2 + \beta)}, \quad B = \frac{2p_3}{(p_3 + q_3) + i(\alpha\underline{p}^2 + \beta)}.$$

It is convenient to express A and B by the energy $E = p^2$ and the angle η, $\operatorname{tg}\eta = |\underline{p}|/p^3$.

We shall also consider $|A|$ and the phase ϕ.

$B = 1 + A$

We get

$$p_3^2 = E\cos^2\eta$$

$$q_3^2 = E\cos^2\eta - V$$

$$|\underline{p}|^2 = E\sin^2\eta \quad .$$

and hence

$$|A| = \left[\frac{(\sqrt{E}\cos\eta - \sqrt{E\cos^2\eta - V})^2 + (\alpha E\sin^2\eta + \beta)^2}{(\sqrt{E}\cos\eta + \sqrt{E\cos^2\eta - V})^2 + (\alpha E\sin^2\eta + \beta)^2}\right]^{\frac{1}{2}}$$

$$\mathrm{tg}\phi = \frac{2\sqrt{E}\cos\eta(\alpha E\sin^2\eta + \beta)}{(\alpha E\sin^2\eta + \beta)^2 - V}$$

We obtain our cases 1),2),3),4) by specification of the constants α,β,V.

Let us discuss the reflection amplitude for very small and very large energies.

	$E \to 0$	$E \to \infty$
1) $H=-\Delta+V(x)$	$\|A\|\to 1$, $\phi=-\pi$	$\|A\|\to 0$, $\phi=-\pi$
2) $H=-\Delta+\delta(x_3)\beta$	$\|A\|\to 1$, $\phi\to 0$	$\|A\|\to 0$, $\phi=\pi/2$
3) $H=-\Delta$ $-\delta(x_3)(\alpha\Delta_2-\beta)$	$\beta>0$, $\|A\|\to 1$, $\phi\to 0$	$\eta=0$ $\|A\|\to 0$ $\phi\to\pi/2$ $\eta\neq 0$ $\|A\|\to 1$ $\phi\to 0$
	$\beta=0$: $\eta=0$, $A\equiv 0$ $\eta\neq 0$, $\|A\|\to 0$ $\phi\to\pi/2$	$\eta=0$, $A\equiv 0$ $\eta\neq 0$, $\|A\|\to 1$ $\phi\to 0$
4) $H=-\Delta+V(x)$ $-\delta(x_3)(\alpha\Delta_2-\beta)$	$\|A\|\to 1$, $\phi\to 0$	$\eta=0$ $\|A\|\to 0$, $\phi\to\pi/2$ $\eta\neq 0$ $\|A\|\to 1$, $\phi\to 0$

These examples show that the additional kinetic energy terms indeed change the character of the reflection and it should be possible to discover the corresponding efect (if it exists) in the experimental data. On the other hand the discussed models may be too simple to describe a real situation.

ACKNOWLEDGEMENT

The author acknowledges gratefully the hospitality at the Research Center BiBoS (Bielefeld) where this work has been done.

REFERENCES

[1] Grossmann A.,Hoegh-Krohn R., Mebkhout, Comm.Math.Phys. 77(1980)87-110.

[2] Karwowski W., Marion J., J.Funct.Analys.62(1985)266-275.

WAVEGUIDES AND CRYSTALS

THIN LATTICES AS WAVEGUIDES

B.S.Pavlov
Department of Mathematical and
Computational Physics, Institute for
Physics, Leningrad State University,
198904 Leningrad, St.Peterhoff, USSR

This lecture is concerned with infinite lattices embedded into a configurational space of a higher dimension. We shall call them thin lattices. Such a structure is non-compact and the scattering on it exibits some pecular features. In particular, the continuous spectrum of related operators consists of two branches :

(S) Scattered wave branch Σ_s: the corresponding eigenfunctions have a form of free waves, reflected by the lattice. This branch coincides with the spectrum of the "unperturbed operator" which acts on the space without any lattice.

(W) Waveguide branch Σ_w: the corresponding eigenfunctions are localised in a neighbourhood of the lattice. In a periodic case they have a form of Bloch waves.

The possible existence of the waveguide branch shows that the system of scattered waves might not be complete in the corresponding Hilbert space,i.e.,that the related wave operators might not be complete. Usually one tries to exclude the waveguide branch, formulating suitable conditions for completeness of the wave operators. On the contrary, we shall discuss here several situations in which this branch plays an important role.

1. One-electron model of a linear molecule

One of the first examples of a thin lattice was studied in 1966-1968 by R.A.Subramanian (see [1]), a postgraduate student of professor Yu.N.Demkov. He suggested a model of a long molecula based on the self-adjoint extension ($-\Delta_\gamma^1$) of Laplace operator with the boundary conditions on the lattice

$$(x_s)^1 = \left\{ x=es,\quad s\in Z^1,\quad |e| = 1,\quad x\in R^3 \right\}$$

$$D(-\Delta_\gamma^1) = \left\{ u:\ u\in W_2^2(R^3\setminus(x_s)^1),\quad u\sim\frac{u_s^1}{4\pi(x-x_s)} + u_s^o + o(1),\right.$$

$$\left. x\to x_s,\quad u_s^o = \gamma\, u_s^1 \right\}.$$

The spectral properties of the operator $(-\Delta_\gamma^1)$ have been investigated by S.Albeverio, R.Høegh-Krohn and others (see [2]). In Ref.3, Yu.E.Karpeshina suggested the following new representation of the related lattice sums

$$D(\lambda,t) = \frac{ik}{4\pi} + \sum_{s\in Z^{(1,2,3)}} \frac{\exp ik|x_s|}{4\pi|x_s|} \exp i(t,x_s)$$

This analysis became a base for more refined crystal models with two-dimensional lattices in R^3, and was used later on for constructing point-interaction models with an internal structure (see [4]); thin Z^1 and Z^2 lattices of zero-range potentials have been investigated in our paper [5]. We begin here describing the results obtained there for Z^1 lattices .In a sense, they are characteristic for all the waveguide situations discussed in the present lecture.

Let $\mathcal{E}^{int} = \oplus\sum E_s$ be an ortogonal sum of unitary equivalent finitedimensional Hilbert spaces. Let A_s be self-adjoint operators in E_s which are mutually unitary equivalent, and $A_{int} = \oplus\sum_{s\in Z^1} A_s$.

The starting operator is defined as a direct sum

$$(-\Delta)\oplus A_{int} \quad \text{in} \quad L_2(R^3)\oplus\mathcal{E}^{int}$$

of the kinetic energy operator $(-\Delta)$ and the "inner" operator A^{int}. The restriction of $(-\Delta)\longrightarrow(-\Delta)_0$ on the linear set D_0^{ext} of all W_2^2 smooth functions in $R^3\setminus(x_s)^1$, which have asymptotics

$$u(x) \underset{x\to x_s}{=} \frac{u_s^1}{4\pi|x-x_s|} + u_s^o + o(1),$$

creates the non-zero boundary form

$$J_0(u,v) = \left\langle (-\Delta)^*_0 u, v \right\rangle - \left\langle u, (-\Delta)^*_0 v \right\rangle = \sum_{s\in Z^1} (u^0_s \overline{v^1_s} - u^1_s \overline{v^0_s})$$

The restriction of the inner operator $A_{int} \to A_{int,o}$ to the linear set D^{int}_0 described in [4,5,6] also leads to nontrivial boundary form $J_1(u^{int}, v^{int})$:

$$J_1(u^{int}, v^{int}) = \sum_{s\in Z} (\xi^-_s(u)\overline{\xi^+_s(v)} - \xi^+_s(u)\overline{\xi^-_s(v)})$$

The boundary form $J_0 + J_1$ of the operator $(-\Delta)_0 + A_{int,o}$ vanishes on the Lagrange plane $L_\Gamma \subset L_2(R^3) \oplus \mathcal{E}_{int}$ given by the translation-invariant boundary conditions

$$\begin{pmatrix} u^0_s \\ \xi^-_s(u) \end{pmatrix} = \sum_{t\in Z^1} \Gamma_{s-t} \begin{pmatrix} u^1_t \\ \xi^+_t \end{pmatrix}, \quad \Gamma_{-t} = \Gamma^*_t ; \; \Gamma_t = 0, |t| > N \qquad (1)$$

or

$$\begin{pmatrix} -u^1_s \\ \xi^-_s \end{pmatrix} = \sum_{t\in Z^1} \Gamma_{s-t} \begin{pmatrix} u^0_t \\ \xi^+_t \end{pmatrix}$$

In both cases the spectral structure of the resulting self-adjoint operators

$$\mathcal{A}_\Gamma = \left\{ (-\Delta)^*_0 \oplus A^*_{int,o} \right\} \Big| D_\Gamma,$$

$$D_\Gamma = \left\{ D^{ext}_0 \oplus D^{int}_0 \right\} + L_\Gamma$$

is similar. In what follows, we shall discuss only the first one.

The spectrum of $\mathcal{A}_\Gamma$ is purely continuous and consist of two branches: the scattered-waves branch Σ_s and waveguide branch Σ_w. The eigenfunctions corresponding to Σ_s have a form of scattered waves:

$$\Psi(x,\nu,k) = \begin{cases} \exp-ik(x,\nu) + \Psi^{ext}_0(k, \;) \sum_{s\in Z^1} \exp-ik(se,\nu) \dfrac{e^{ik|x-se|}}{4\pi|x-se|} \\ \exp-ik(se,\nu)\, \Psi^{int}_0(k,\nu) \end{cases}$$

for $k=\sqrt{\lambda}$ and $\lambda \in [0,\infty)$. The eigenfunctions corresponding to Σ_w have a form of Bloch-waves localised near the lattice :

$$\varphi(x,t,k) = \begin{cases} \sum\limits_{s\in Z^1} \exp -ist \dfrac{\exp ik|x-se|}{4\pi|x-se|} \\ \exp -ist\, \varphi_0(k) \end{cases}$$

The energy λ and the quasimomentum t of the waveguide branch are connected by the dispersion equation

$$\det \left\{ \begin{pmatrix} d(\lambda) & 0 \\ 0 & D(\lambda,t) \end{pmatrix} - \sum_{s\in Z} \Gamma_s \exp -its \right\} = 0,$$

where the R-function $d(\lambda) = \langle (I+\lambda A)(A-\lambda I)^{-1}\theta,\theta \rangle$ depends on the spectral structure of the inner operator $A=A_s$ and the deficiency vector θ of its restriction in E_s. The lattice-sum $D(\lambda,t)$, corresponding to the one-dimensional lattice $\{x_s\}$ was calculated explicitely by R.A.Subramanian :

$$D(\lambda,t) = \frac{1}{4\pi} \ln(2(\cos\sqrt{\lambda} - \cos t))^{-1}$$

The branch of logarithm is fixed by the condition of analytical continuability of $D(\lambda,t)$ into the complex spectral plane of λ and real-valuedness of $D(\lambda,t)$ on the negative semiaxis. In the case without "atomic orbitals overlaping" ,i.e., if $\Gamma_s=0$ for $s\neq 0$ and

$$\Gamma_0 = \begin{pmatrix} \Gamma_{00} & \Gamma_{01} \\ \Gamma_{10} & \Gamma_{11} \end{pmatrix}$$

then the dispersion equation becomes more simple

$$(\Gamma_{00} - D(\lambda,t))(\Gamma_{11} - d(\lambda)) - |\Gamma_{01}|^2 = 0 \tag{2}$$

If the interaction between the internal and external channels is weak, $|\Gamma_{01}| \ll 1$, this equation can be easily solved graphicaly. In this case the zero-lines of eq.(2) in the (t,k)-plane, $k=\sqrt{\lambda}$, are approximately given by the zero lines of $\Gamma_{00} - D = 0$ and $\Gamma_{11} - d = 0$. Our choice of the logatithm branch in $D(\lambda,t)$ yields the equality $\operatorname{Im} D(\lambda,t) = n/4$, $n\in Z^1$, on the domains cut from the strip $\{t: -\pi < t < \pi\}$ by the lines $k\sqrt{\lambda} \pm t = 2\pi z$, $z\in Z^1$. The real part of $D(\lambda,t)$ has logarithmic singularities on these lines. The waveguide branch Σ_w found by R.A.Subramanian is given by projection of the zero-line $\Gamma_{00} - D = 0$ onto the λ-axis. The positive part of this zero-line is denoted as α_1 on Fig.1. The waveguide branch Σ_w , which corresponds to the operator A is given by projection of the lines α_2(see on the left of Fig.1). The quasistationary bands corresponding to higher momenta n=1,2,... have been treated in Ref.7. In our model these bands arise from the inner structure of the atoms,i.e., the internal structure of the zero-range potentials.

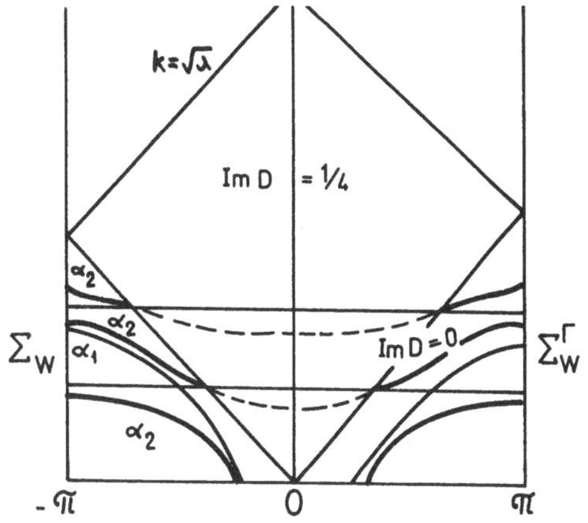

Fig.1

This nontrivial structure of the "atoms" considered in our model illustrates the relation between the single-atom resonances which are localised near the zero lines $\Gamma_{11} - d(\lambda) = 0$ (see [6]) and the gap-band structure of Σ_W One can also evaluate the effective mass of the corresponding quasiparticle $(d^2\lambda / dt^2)^{-1}$. This mass appears to be very large near the edges of the resonant bands and/or gaps. The fact that the waveguide band is situated in the lower part of the spectrum, $-\pi < t < \pi$, corresponds to the requirement that the component of the three-dimensional momentum $k\cos(e,\nu) = \sqrt{\lambda}\ (e,\nu)$ tangential to the lattice is majorized by the lattice quasimomentum t. Thus the "conservation law" for momentum prohibites emission of an electron from the lattice into the R^3 - space.

It is easy to include a finite number of impurities into the lattice assuming

$$A_s = A \quad \text{for} \quad s \in (s_1, s_2, s_3, \dots, s_n) \text{ and}$$

$$A_s \neq A \quad \text{for} \quad s \notin (s_1, s_2, s_3, \dots, s_n)$$

Using the following ansatz for waveguide eigenfunctions

$$\begin{cases} \varphi^{ext}(x,k) = \sum\limits_{s \in Z} \dfrac{\exp ik|x-se|}{4\pi|x-se|} \varphi_s^{ext} \\ \\ \varphi_s^{int}(k) = \widetilde{\varphi} + \xi_s^- W_s^- + \xi_s^+ W_s^+ \end{cases}$$

we get for the boundary values in the exterior space

$$\xi_s^{ext-} = \frac{ik}{4\pi} \varphi_s^{ext} + \sum_{r \neq s} \frac{\exp ik|s-r|}{4\pi|s-r|} \varphi_r^{ext}$$

$$\xi_s^{+ext} = \varphi_s^{ext}(k)$$

$$\xi_s^{-int} = \left\langle \frac{I - \lambda A_s}{A_s - \lambda I}\,\theta,\ \theta \right\rangle \xi_s^{+int} = d_s(\lambda)\,\xi_s^{+int}$$

The boundary conditions (1) give the equation

$$\begin{pmatrix} \dfrac{ik}{4\pi}\varphi_s^{ext} + \sum\limits_{r\neq s} \dfrac{\exp ik|s-r|}{4\pi|s-r|}\varphi_r^{ext} \\ \xi_s^{+int} \end{pmatrix} = \begin{pmatrix} \Gamma_{oo} & \Gamma_{o1} \\ \Gamma_{1o} & \Gamma_{11} \end{pmatrix} \begin{pmatrix} \varphi_s^{ext} \\ \xi_s^{+int} \end{pmatrix}$$

which can be solved easily in terms of Fourier transforms of φ_s^{ext} and ξ_s^{+int}

$$\varphi(q) = \sum_{r\in Z} \exp iqr\ \varphi_r^{ext},\quad \xi(q) = \sum_{r\in Z} \exp iqr\ \xi_r^{+int},\ -\pi<q<\pi$$

and the lattice sum $D(\lambda, q)$

$$\begin{pmatrix} D(\lambda,q)\,\varphi(q) \\ d(\lambda)\,\xi(q) + (K\xi)(q) \end{pmatrix} = \begin{pmatrix} \Gamma_{oo} & \Gamma_{o1} \\ \Gamma_{1o} & \Gamma_{11} \end{pmatrix} \begin{pmatrix} \varphi(q) \\ \xi(q) \end{pmatrix} \qquad (3)$$

Here K is the finite-rank integral operator in $L_2(-\pi, \pi)$ with the kernel

$$K(q-p) = \sum_{s\in(s_1,s_2,\dots,s_n)} (\,d_s(\lambda) - d(\lambda)\,)\exp is(q-p)$$

Excluding φ from the last equation,

$$\varphi(q) = -\,(\,\Gamma_{oo} - D(\lambda,q)\,)^{-1}\,\Gamma_{o1}\,\xi(q)$$

we get the Friedrichs-model-type equation for ξ

$$(\,(\Gamma_{11} - d(\lambda)) - \Gamma_{1o}(\,\Gamma_{oo} - D(\lambda,q)\,)^{-1}\,\Gamma_{o1})\xi - K\xi = 0 \qquad (4)$$

The simplest equation of this type with a smooth kernel K was investigated by L.D.Faddeev [8] . By a similar argument, one can show that the waveguide branches of spectrum for the homogeneous and inhomogeneous lattice coincide. The only difference is that some additional (finitely many) eigenvalues and resonances can appear in the inhomogeneous-lattice case. The operators in the lhs of eq.(4)

are comparable for k=0 and K≠0 independently of λ. The corresponding scattering matrix which describes scattering processes in the waveguide can be calculated explicitly.

The case of a halflattice, $d_s=0$ for $s<0$, is also explicitly solvable. The role of the integral operator K is then played by a Henkel operator. One can combine the above discussed cases and discuss half-lattices with impurities. Other posibilities such as $A_s \to A_{\pm}$ for $s \to \pm\infty$ are interesting rather mathematically than from the physical point of view. A similar consideration can be performed for a two-dimensional lattice in R^3. The homogeneous lattice $\{z_s\}^2$, $s \in Z^2$ is considered in [5].

2. One-electron model of a long elastic molecule

We take the simplest one-particle Hamiltonian (see also [9]) in the form of a Jacobi matrix H_{ph} in $\ell_2(Z^1) = H^1_{ph}$

$$H^1_{ph} = \gamma^1_{-1} T_1 + \gamma^1_o T + \gamma^1_1 T^+_1$$

where $(T_1 u)_s = u_{s-1}$, $u \in \ell_2(Z^1)$. The N-phonon Hamiltonian in the harmonic approximation can be considered as the form sum

$$H_{ph} = H^1_{ph} \otimes I^2 \otimes \ldots \otimes I^N + I^1 \otimes H^2_{ph} \otimes I^3 \otimes \ldots \otimes I^N + I^1 \otimes I^2 \otimes \ldots \otimes H^N_{ph}$$

It is a selfadjoint operator in Hilbert space $\mathcal{H}_{ph} = \mathcal{H}^1_{ph} \otimes \mathcal{H}^2_{ph} \otimes \ldots \otimes \mathcal{H}^N_{ph}$.

The simplest one-particle Hamiltonian of a long molecule can be taken in the form of a one-dimensional Schrodinger operator with a periodic point potential (the Kronig-Penney model) : in $L_2(R) = \mathcal{H}_{el}$ we define

$$H\,u = \left(- \frac{d^2}{dx^2} + \sum h\,\delta(x-n) \right) u, \qquad u \in W^2_2(R \setminus Z) \cap W^1_2(R).$$

It is easy to combine H_{ph} and H_{el} to construct the initial Hamiltonian

$$H_{el} \otimes I_{ph} + I_{el} \otimes H_{ph} = H$$

which is a selfadjoint operator in the tensor product $\mathcal{H} = \mathcal{H}_{el} \otimes \mathcal{H}_{ph}$

The spectrum of H is the algebraic sum of the spectra σ_{el} of H_{el} and σ_{ph} of H_{ph}. If p_{el} is the quasimomentum of H_{el} and X(p) is the corresponding eigenfunction :

$$H_{el}\, X(p_{el}) = \alpha_{el}(p_{el})\, X(p_{el})$$

and $X(p_{ph})$ is the phonon eigenfunction with quasimomentum p_{ph}

$$H_{ph}\, X(p_{ph}) = \alpha_{ph}(p_{ph})\, X_{ph} ,$$

then $X(p) = X(p_{el}) \cdot X(p_{ph})$ is the eigenfunction of H with the quasimomentum $p = (p_{el}, p_{ph})$ corresponding to the eigenvalue

$$\alpha(p) = \alpha_{el}(p_{el}) + \alpha_{ph}(p_{ph})$$

The resolvent of H can be easily constructed in terms of its eigenfunction expansion. Its kernel

$$G\left(\begin{pmatrix} x \\ s \end{pmatrix}, \begin{pmatrix} y \\ r \end{pmatrix}, \lambda\right) = \int \frac{X(p,{}^{x}_{s})\; X(p,{}^{y}_{r})}{\alpha(p) - \lambda}\, dp$$

is a function on the electron-phonon configuration space which is the union of a countable family of copies of the real axis Λ_s

$$R \times Z^N = \sum_{s \in Z^N} \Lambda_s .$$

It fulfils the equation

$$(H - \lambda I)\, G({}^{x}_{s},{}^{y}_{r},\lambda) = (\ldots,0,\delta_{sr}\,\delta(x-y),0,\ldots)$$

If y is integer, y=m, then the additional singularity of G can be described in the boundary-condition form :

$$[G_r'](n) - h\, G_s(n) = -\delta_{nm}\,\delta_{sr} ,$$

here $G_s = G \restriction \Lambda_s$. We interpret s as the coordinate of an excited oscillator. Then the interaction between phonons and electrons must be localized near the main diagonal

$$M = \left\{ x=m,\ s=(m,m,m,\ldots,m) \right\}$$

of the configuration space. The simplest way to switch in the interaction is given by the selfadjoint-extension theory. Let us restrict $H \to H_0$ to the operator defined on the linear set D_0 of all functions u, v $\in D(H)$ vanishing on the main diagonal. An integration by parts leads to the boundary form

$$\langle H_0^* u, v \rangle - \langle u, H_0^* v \rangle = \sum_{m \in Z^1} \Big\{ ([u_m'](m) - h\, u_m(m))\, \overline{v_m(m)} - - u_m(m)\, \overline{([v_m'](m) - h\, v_m(m))} \Big\}$$

Using the boundary conditions on the diagonal M

$$[u_m'](m) - h\, u_m(m) = \Gamma u_m(m), \quad \Gamma \neq 0, \quad \operatorname{Im} \Gamma = 0 \tag{4}$$

we get a homogeneous one-dimensional Hamiltonian H_Γ. Of course, one can use also boundary conditions depending on energy, $\Gamma \to \Gamma(\lambda)$, or to employ the next-to-main diagonals. Such assumptions also lead to solvable models. However, we restrict ourselves here to the simplest sort of boundary conditions (4). The spectrum of H consists of two branches Σ_s and Σ_w. The scattering branch Σ_s coincides with $\sigma(H)$. The corresponding scattered waves differ from the electron-phonon Bloch functions by the waves reflected from the diagonal, which plays here the role of a thin lattice

$$\Psi_p = X_p + \sum \Psi_m G_m, \quad G_m = G(\tfrac{x}{n}, \tfrac{m}{m}, \lambda).$$

Using the translation invariance, we get $\Psi_m = \Psi_0 \exp i(p_{el} + p_{ph})m$ and

$$\Psi_0 = - X(p, \tfrac{0}{0}) \Big(\frac{1}{r+h} + \sum_{m \in Z} \exp im(p_{el}+p_{ph})\, G(\tfrac{0}{0}, \tfrac{m}{m}, \lambda + i0) \Big).$$

The scattering amplitude can be derived from this formula using the asymptotic behaviour at infinity, $(x,s) \to \nu\infty$, $\nu \in R^2$, $|\nu| = 1$

$$G(\tfrac{x}{s}, \tfrac{y}{n}, \lambda) \sim \overline{X_{p^*}(\tfrac{y}{n})}\, G(\tfrac{X}{s}, \tfrac{0}{0}, \lambda) \tag{5}$$

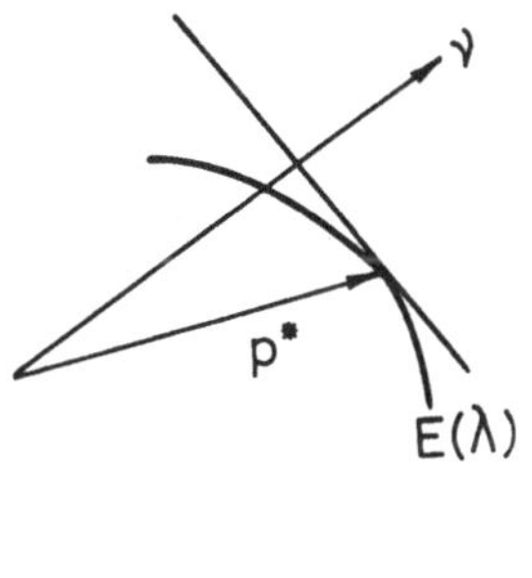

Here $p^* = (p_{el}^*, p_{ph}^*)$ is the point on the isoenergetic surface E_λ, where the tangent line is orthogonal to ν. We assume here, that there exist only one point of such a sort (the formula (5) has been derived by my student V.Evstratov). The asymptotics (5) gives at infinity

$$\Psi_p(\tfrac{x}{s}) \sim X_p(\tfrac{x}{s}) - \frac{2\pi X_p(\tfrac{0}{0}) \overline{X_{p^*}(\tfrac{0}{0})} \sum_{z \in Z} \delta(p_{el}+p_{ph}-p_{el}^*-p_{ph}^*+2\pi r)}{(\Gamma+h)^{-1} + \sum_{m \in Z} \exp im(p_{el}+p_{ph})\, G(\tfrac{0}{0}, \tfrac{m}{m}, \lambda+i0)}\, G(\tfrac{x}{s}, \tfrac{0}{0})$$

These functions describe the processes of electron-phonon scattering. The corresponding amplitude contains Laue singularities and has, due to the denominator, a resonant form. The second branch Σ_W^{Γ} corresponds to the eigenfunctions which are localised near the diagonal M and have the form of Bloch waves :

$$\varphi_q = \sum_{m \in Z^1} \exp i(q,m)\, G_m, \quad m = (m,m,\dots,m)$$

The quasimomentum q can be calculated from the following dispersion equation

$$(h + \Gamma)^{-1} + \sum_{m \in Z^1} \exp i(q,m)\, G(\begin{smallmatrix}0 & m\\ 0, & m\end{smallmatrix}, \lambda + i0) = 0 \qquad (6)$$

This equation is similar to the equation (2) of Section 1. The corresponding quasiparticles can be interpreted as polarons.

Using energy-dependent boundary conditions with an R-function $h=h(\lambda)$, we get the picture of waveguide bands, which is quite similar to the corresponding picture of section 1. It is not difficult to discuss the model of a crystal built of atoms of different sorts. There are no difficulties to consider crystals in R^3 and crystals with finite number of impurities. More difficult problems arise for semiinfinite lattices and for lattices which contains stochastic elements. If those elements are given by Markovian processes, then the electron-phonon properties of the corresponding lattices can be studied after averaging which yields nonhermitian Hamiltonians.

Of course, one can impose boundary conditions not only on the main diagonal, but also on the diagonal planes

$$M_{i_1 \dots i_n} = \left\{ (x,s):\ x=m,\ s_i=m,\ m \in Z^1,\ i \in (i_1,i_2,\dots,i_n),\ n < N \right\}$$

Then we get a many-body problem with many channels. It can be studied with the help of appropriate technical tools (see [9]). In this model, one can describe the scattering of polarons of different sorts. These models are not solvable, but equations, which appear in corresponding analysis are quite similar to Faddeev's equations (see the zero-range version of them in [10, 11]).

3. Two quantum particles in a one-dimensional periodic lattice

This is, in fact, a three-body quantum problem (see also [12]) but it can be reduced to a one-dimentional integral equation which gives again the waveguide-type eigenfunctions.

We construct a one-dimentionalhard lattice of zero-range potentials with an internal structure. The one-dimentional one-particle Hamiltonian is a selfadjoint extention of the restricted "original" Hamiltonian

$$\mathcal{A}_0 = \left\{ -\frac{d^2}{dx^2} \oplus \sum_{n\in Z^1} (A)_n \right\}_0 = \left\{ -\frac{d^2}{dx^2} \oplus A^{in} \right\}_0$$

in $$\left\{ L_2(R) \oplus \sum_{n\in Z^1} (E)_n \right\} = L_2(R) \oplus \mathcal{E}^{in} = \mathcal{H}$$

All the internal operators are unitary equivalent, $(A)_n \sim A$, and the deficiency elements θ_s of the corresponding restrictions coincide mutually, $\theta_s=\theta_r=\theta$. The external operator is restricted by the condition of vanishing at $x=s$. Then (see [4])

$$D(A_0)_s = \left\{ u_s:\ u_s = \frac{1}{A-iI} U^{\perp} + \xi_s^+ W_s^+ + \xi_s^- W_s^- \right\},$$

$$D\left(-\frac{d^2}{dx^2}\right)^* = \left\{ u:\ u \in W_2^2(R\setminus Z) \cap C \right\}.$$

We choose the extension $\mathcal{A}_\Gamma$ of $\mathcal{A}$ determined by the following boundary conditions

$$\begin{pmatrix} -u(s) \\ \xi_s^- \end{pmatrix} = \Gamma \begin{pmatrix} [u'](s) \\ \xi_s^+ \end{pmatrix}, \quad \Gamma = \Gamma^*,$$

which gives a Schrödinger equation with energy-dependent potential for the external components of eigenfunctions

$$-\frac{d^2}{dx^2} u(x) - \left\{ \Gamma_{00} - |\Gamma_{01}|^2 (\Gamma_{11} - d(\lambda))^{-1} \right\} \sum_{n\in Z^1} \delta(x-n) u(x) =$$

$$= \lambda\, u(x) ;$$

here $d(\lambda) = \left\langle (A-\lambda I)^{-1} (1+\lambda A)\, \theta, \theta \right\rangle$. The spectrum of A_Γ can be derived from the dispersion equation

$$(\Gamma_{oo} - D(\lambda,t)) (\Gamma_{11} - d(\lambda)) - |\Gamma_{1o}|^2 = 0$$

Here $D(\lambda,t)$ is one-dimensional lattice sum,

$$D(\lambda,t) = (2\sqrt{\lambda})^{-1} \sin\sqrt{\lambda} \; (\cos\sqrt{\lambda} - \cos t)^{-1}.$$

In order to construct the two-body Hamiltonian we can start with the operator $A^1 \otimes I^2 + I^1 \otimes A^2$ in the space $\mathcal{H}_1 \otimes \mathcal{H}_2$. The decomposition

$$\mathcal{H}^1 \otimes \mathcal{H}^2 = L_2(R_1 \otimes R_2) \oplus \{L_2(R_1) \otimes \mathcal{E}_2\} \oplus \{\mathcal{E}_1 \otimes L_2(R_2)\} \oplus \{\mathcal{E}_1 \otimes \mathcal{E}_2\}$$

contains the subspace $\mathcal{E}_1 \otimes \mathcal{E}_2$ which can be neglected if we discuss only the interaction of "free" particles with the lattice. Nevertheless, it must be taken into account if we want to consider the direct interaction between the internal dynamics of the "atoms" A_s, A_t. We begin with the "original" two-body Hamiltonian

$$\mathcal{A} = (- \frac{d^2}{dx_1^2} - \frac{d^2}{dx_2^2}) \oplus (A_1^{int} \otimes I_{x_2} + I^1 \otimes (- \frac{d^2}{dx_2^2})) \oplus$$

$$\oplus (I_{x_1} \otimes A_2^{int} + (- \frac{d^2}{dx_1^2}) \otimes I^2) \oplus \sum_{s,t \in Z^1} ((A)_s \otimes I_t + I_s \otimes (A)_t \tag{7}$$

Here $A^{int} = \oplus \sum_{s \in Z^1} (A)_s$ and I is the identity in $\mathcal{E}$. The interaction between "free" particles can be described by a potential in the first term, which is then replaced by a Schrodinger operator,

$$- \frac{d^2}{dx_1^2} - \frac{d^2}{dx_2^2} \longrightarrow - \frac{d^2}{dx_1^2} - \frac{d^2}{dx_2^2} + q(|x_1-x_2|) .$$

Here q can be Coulomb or Yukawa potential or even some separable interaction in $L_2(R^2) = L_2(0,\infty) \otimes L_2(0,2\pi)$. The lattice effects can be introduced by the above described restriction-extension procedure. The simplest two-body interaction can be also taken in the form $h(x_1-x_2)$, which is equivalent to the boundary condition $\left[\frac{\partial u}{\partial n} \right] = hu$ on the diagonal M of the configuration space R^2.

Unfortunately, the last condition is not valid for fermions, which have antisymmetric eigenfunctions. Since they are continuous they vanish on the diagonal M and the last boundary condition becomes trivial. To rectify this situation we use futher a smooth potential and resctrict the external operator to the linear set of all smooth functions, vanishing on the lines $L_s^l = \{x_l = s, l=1,2, s \in Z^I\}$. The restriction of the inner operators

$$A_{1s}^{int} \otimes I_{x_2} + I_{1s} \otimes \left(-\frac{d^2}{dx_2^2}\right), \quad I_{x_1} \otimes A_{2t}^{int} + \left(\frac{d^2}{dx_1^2}\right) \otimes I_{2t},$$
$$(A_s \otimes I_t + I_s \otimes A_t)$$

gives a nontrivial boundary form, written in terms of the boundary values $\xi_{1s}^{\pm}(x_2), \xi_{2t}^{\pm}(x_I), \xi_{st}^{\pm}$ as

$$\begin{aligned}
&\sum_{\substack{l=1,2\\ s\in Z}} \int_{L_s^l} \left\{\left[\frac{\partial u}{\partial n}\right]\bar{v} - u\overline{\left[\frac{\partial v}{\partial n}\right]}\right\} dx_l + \sum_{s,t} \left\{ u_{st}^-(u)\overline{u_{st}^+(v)} - \right. \\
&\left. - u_{st}^+(u)\overline{u_{st}^-(v)} \right\} + \sum_{s \in Z} \int_{L_s^1} \left[\xi_{1s}^-(u)\overline{\xi_{1s}^+(v)} - \right. \\
&\left. - \xi_{1s}^+(u)\overline{\xi_{1s}^-(v)} \right] dx_1 + \sum_{t \in Z} \int_{L_t^2} \left[\xi_{2t}^-(u)\overline{\xi_{2t}^+(v)} - \right. \\
&\left. - \xi_{2t}^+(u)\overline{\xi_{2t}^-(v)} \right] dx_1 + \sum_{s,t} \left[\xi_{st}^-(u)\overline{\xi_{st}^+(v)} - \xi_{st}^+(u)\overline{\xi_{st}^-(v)} \right].
\end{aligned} \tag{8}$$

Here the second term corresponds to the deficiency elements which have singularities at the intersections of the lines $L_s^1 \cap L_t^2$. It can be shown that every intersection contributes by one deficiency indices. The last term corresponds to the purely internal term in (7). The boundary form (8) vanishes on the linear set D_Γ of all elements u, $u \in D_0^*$, which are subjected to the boundary conditions with hermitian 2×2 matrices Γ, Γ_{st}

$$\begin{pmatrix} -u(s,x_1) \\ \xi_{1s}^-(x_1) \end{pmatrix} = \Gamma \begin{pmatrix} \frac{\partial \varphi}{\partial n}(s,x_1) \\ \xi_{es}^+(x_1) \end{pmatrix}, \qquad (s,x_1) \in L_s^1$$
$$\begin{pmatrix} u_{st}^- \\ \xi_{st}^- \end{pmatrix} = \Gamma_{st} \begin{pmatrix} u_{st}^+ \\ \xi_{st}^+ \end{pmatrix}, \qquad s,t \in Z^I \tag{9}$$

As far as the overlapping of the "orbitals" is nontrivial for the neighbouring points only, we put $\Gamma_{st} = \Gamma_{s-t} = 0$ for $|s-t| > 1$.

It can be shown that the operator $A_0 \mid D_\Gamma$ is semibounded from below. Its Friedrichs extensions plays the role of the two-body Hamiltonian in the periodic lattice. To construct the resolvent of this Hamiltinian, one has to solve the system of the equations

$$\begin{aligned} & -\frac{d^2}{dx_1^2} u - \frac{d^2}{dx_2^2} u + q(|x_1 - x_2|)\, u - \lambda u = f_0 , \\ & \left\{ (A_s)_0^* - (\lambda + \frac{d^2}{dx_2^2}) \right\} u_1^s = f_1^s , \\ & \left\{ (A_t)_0^* - (\lambda + \frac{d^2}{dx_1^2}) \right\} u_2^t = f_2^t , \\ & (A_s + A_t)_0^* u_{st} - \lambda u_{st} = f_{st} , \qquad s,t \in Z^I \end{aligned} \tag{IO}$$

with the boundary conditions (9).

In the case when the direct overlapping of orbitals is neglected the last terms in (7),(8),(9), $s,t \in Z^I$, must be dropped. Then the the system (10) can be reduced (see Ref.10 for details) to the Schrödinger equation in the external channel with the energy-dependent additional potential containing pseudo-differential operators on the lines:

$$-\frac{d^2}{dx_I^2} u - \frac{d^2}{dx_2^2} u + q(x_1 - x_2)\, u +$$

$$+ \sum_{\substack{l=1,2 \\ s}} \delta(x_l - s) \left\{ \Gamma_{oo} - |\Gamma_{o1}|^2 \left[\Gamma_{11} - d(\lambda + \frac{d^2}{dx_1^2}) \right] \right\}^{-I} u = \lambda \varphi + f .$$

In the case when the internal operator $A = \sum \alpha_s E_s$ is of a finite rank the pseudo-differential operator $d(\lambda + \frac{d^2}{dx^2})$ has the following form:

$$d(\lambda + \frac{d^2}{dx^2}) = -I - \int_{-\infty}^{\infty} \sum_s \frac{I + \alpha_s^2}{2i\sqrt{\alpha - \alpha_s}} \langle E_s \theta, \theta \rangle_e\, e^{i\sqrt{\alpha - \alpha_s}\, |x-t|} dt .$$

Finally, we derive the dispersion equation for the waveguide branch Σ_W of the spectrum. We assume that $q < 0$ and insert a new function in the exterior channel, $v = q \cdot u$. Using then the exterior quasiresolvent G corresponding to $q = 0$, we get the equation for v:

$$v(x) = \int_{R^3} \sqrt{q(x)}\; G_\lambda(x,y)\sqrt{q(y)}\; v(y)dy ,$$

Here $q(x) = q(|x_1 - x_2|)$.

Taking into account the Bloch property of the waveguide eigenfunction with respect to the shift along the lattice $x \to x+e$, $e=(1,1)$, we can reduce the last integral equation to a simpler one with the operator K which is compact in the Yukava case:

$$v(x) = \int_{0 \leq y_1+y_2 \leq 2} \sqrt{q}\,(x)\; K_\lambda(x,y,p)\sqrt{q}\,(y)\; v(y)dy\ ,\quad 0 \leq x_1+x_2 \leq z\ ,$$

whose kernel is the-lattice sum:

$$K_\lambda(x,y,p) = \sum_{m \in Z} G_\lambda(x,y+me)\; e^{ipm} .$$

Thus the dispersion equation for waveguide branch Σ_w can be written in terms of a Fredholm determinant containing the potential and the quasiresolvent of the initial two-particle Hamiltonian without interaction ($q=0$). This quasiresolvent can be calculated explicitly.

It is very interesting to investigate the localization of the waveguide bands. This can be done numericaly on a computer.

The waveguide bands can exist even in the case of a repulsive potential, $q>0$, because the effective mass of a particle is negative. One can also construct more realistic models using the electron-phonon blocks described in previous sections as initial operators. This will be done in another publication.

In any case, we get different kinds of waveguide bands, which correcpond to forming and dissociating of the combined quasiparticles similar to bipolarons. More difficult analytical problems arise in the case of many particles, when the bound states of three, four etc. electrons or polarons could appear.

One can also insert the interaction between different phonon modes in our modell. To do it one takes as usually a harmonic term in potentials. The restriction-extension procedure gives another possibility to switch in the interaction between different phonon modes, based on a construction similar to that described in previous section. It will be done in another publication.

The approach to the description of interactions suggested here looks at glance very different from the standard one. In fact, it is very close to it provided there are invariant subspaces of the standard Hamiltonian which contain a finite number of particles only. Moreover, using the internal structure we are able to enrich the spactral properties of the models without any analytical difficulties.

It could provide physicists with many attractive possibilities of heuristic considerations based on easily manageable formulae.

References

1. Yu.N.Demkov, V.N.Ostrovsky: The method of zero-range potentials in atomic physics. Leningrad State University Publ. in Russian
2. S.Albeverio, F.Gesztesy, R.Hoegh-Krohn, H.Holden: Solvable models in quantum mechanics. Springer Verlag, Berlin 1988
3. Yu.E.Karpeshina: Teor.Mat.Fiz. 57 (1983) 414 - 423
4. B.S.Pavlov: Teor.Mat.Fiz. 72 (1983) , No 3
5. P.B.Kurasov, B.S.Pavlov: Teor.Mat.Fiz. to appear
6. B.S.Pavlov: Teor.Mat.Fiz 59 (1984) 345 - 354
7. G.V.Golubkov, F.I.Dalidchick, G.K.Ivanov: Sov.Phys. JETP 78 (1980) 1423 - 1434
8. L.D.Faddeev: Proc.of Steklov Inst. 73 (1964) 292 - 313
9. S.P.Merkuriev, L.D.Feddeev: Quantum Scattering Theory For Few-Body Systems. Nauka, Moscow 1986 in Russian
10. B.S.Pavlov: Leningrad Univ.Vestnik, Ser.Mat. 1987
11. Yu.A.Kuperin, K.A.Makarov, S.P.Merkuriev, B.S.Pavlov, A.K.Motovolov: Feddeev equations with enegry.dependent potentals. Preprint ISSN 013-426Y, ITP Budapest, November 1986
12. B.S.Pavlov: Sov.Math.Sbornik to appear
13. L.D.Faddeev, F.A.Berezin: DAN USSR 137 (1961) 1011 - 1014

QUANTUM WAVEGUIDES

P.Exner, P.Šeba, P.Šťovíček
Laboratory of Theoretical Physics,
JINR, 141980 Dubna, USSR

This lecture is concerned with the motion of a quantum particle on stripes, tubes, layers and similar manifolds in $\mathbb{R}^n$. A study of this problem might have started at every moment of the more-than-sixty-years-long history of quantum mechanics. It did not, however, apart of some trivial textbook examples, because an appropriate physical motivation was lacking.

This situation has changed with the advent of modern microelectronics. During recent years, various technologies have been developed which allow to produce ultrathin layers and "wires" or extremely pure metallic or semiconductor materials and build more complicated structures such as sandwiches, graphs with many branching points etc. of them[1]. Our knowledge of quantum-mechanical effects in such systems is at present very poor.

Let us notice that similar problems for a constrained wave equation have been extensively studied in the classical waveguide theory. The electron motion in the above mentioned structures is governed by the Schrödinger equation (with appropriate boundary conditions), so it is natural to call these systems quantum waveguides (or briefly, Q-guides). The analogy is not only illustrative but useful too ; recall that in the stationary approach, the free Schrödinger and wave equations are the same up to physical meaning of the spectral parameter.

1. Bound states in curved Q-guides

We shall consider a free spinless particle living on a curved planar strip Ω of a width d with Dirichlet boundary conditions (Fig.1). It can model the electron motion (with spin neglected) on a planar quantum wire or a thin layer on a cylinder-shaped substrate. Since the particle is assumed to be free, the results

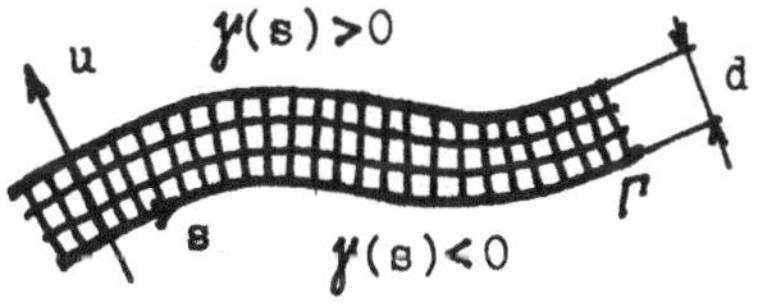

Fig.1

are applicable provided the region of interest is smaller than the mean free path of electrons ; it can be achieved. A more complete description should include, however, a stochastic potential corresponding to random impurities.

We choose one of the boundaries of Ω as the reference curve Γ and introduce the natural curvilinear coordinates s,u on Ω. The strip is then fully characterized by its width d and the (signed) curvature $\gamma(s)$ of Γ.

The state Hilbert space of the problem is $L^2(\Omega, dx\,dy)$ and the Hamiltonian is

$$H_\Omega = -\frac{\hbar^2}{2m}\Delta_D \quad , \tag{1.1}$$

where Δ_D denotes the Dirichlet Laplacian relative to Ω. We shall restrict our attention to the strips with an infinitely smooth boundary which are curved essentially within a bounded region only, i.e., we adopt the following assumptions

(a) $\gamma \in C^\infty$,

(b) $\gamma, \gamma', \gamma''$ are bounded,

(c) $\gamma(s) \geq \gamma_-$ for some $\gamma_- \in (-d^{-1}, 0]$,

(d) $\gamma^2, \gamma'' \in L(\mathbb{R}, (1+|s|)\,ds)$,

(e) $\gamma, \gamma' \in L^2(\mathbb{R}, (1+s^2)\,ds)$.

We shall consider the non-trivial case ($\gamma \neq 0$) only.

The following theorems are proven in Ref.2 . First we pass to a straight-strip problem using the coordinates s and u :

1.1 <u>Theorem</u> : Assume (a). Then H_Ω is unitarily equivalent to a self-adjoint operator H on $L^2(\mathbb{R} \times [0,d])$ such that

$$D(H) \supset D \equiv \{\psi : \psi \text{ is } C^\infty, \psi(s,0) = \psi(s,d) = 0 \text{ for } s \in \mathbb{R}, H\psi \in L^2\}$$

and

$$H\psi = -\frac{\hbar^2}{2m}\left\{\frac{\partial}{\partial s}(1+u\gamma)^{-2}\frac{\partial\psi}{\partial s} + \frac{\partial^2\psi}{\partial u^2}\right\} + V(s,u)\psi \tag{1.2a}$$

for $\psi \in D$, where

$$V(s,u) = \frac{\hbar^2}{2m}\left\{-\frac{\gamma^2}{4(1+u\gamma)^2} + \frac{u\gamma''}{2(1+u\gamma)^3} - \frac{5}{4}\frac{u^2\gamma'^2}{(1+u\gamma)^4}\right\} . \tag{1.2b}$$

A simple minimax estimate applied to the operator (1.2) now gives

1.2 Theorem : Let Ω fulfil (a)-(d) . Then

(i) $\sigma_{ess}(H_\Omega) = [E_\infty, \infty)$, where $E_\infty = \hbar^2\pi^2/2md^2$,

(ii) there is $d_0 > 0$ such that for all $d < d_0$, the operator H_Ω has at least one bound state in $[0, E_\infty)$.

The same argument yields a lower bound on d_0 , but a relatively poor one. A slightly modified Birman-Schwinger technique can be used to prove

1.3 Theorem : Let Ω fulfil (a)-(e), then a bound state $E \in [0, E_\infty)$ exists iff

$$\int_{\mathbb{R}\times[0,d]} \left\{ -\frac{\gamma^2}{(1+u\gamma)^2} + \frac{u^2\gamma'^2}{(1+u\gamma)^4} \right\} \sin^2\left(\frac{\pi u}{d}\right) ds\, du < 0 \quad . \tag{1.3}$$

In particular, if Ω is simply bent, i.e., $\gamma(s) \geq 0$ for all $s \in \mathbb{R}$, then

$$d_0 \geq d_+ \equiv \frac{1}{2\gamma_+}\left(\sqrt{1 + 4\gamma_+\sqrt{z}} - 1\right) \quad , \tag{1.4}$$

where $\gamma_+ = \max_s \gamma(s)$ and $z = \int \gamma^2 ds \Big/ \int \gamma'^2 ds$.

Simple examples show that the estimate (1.4) yields the critical width of the same order of magnitude as the minimum curvature radius, unless γ exhibits steep changes.

The above results say nothing, however, about the gap between the bound state E and the bottom of $\sigma_{ess}(H_\Omega)$ given by the first transversal-mode energy. With this aim, we present here the results of a solvable model. It concerns the rectangular Q-guide (Fig.2). Since Ω is self-similar in this case, we have

$$E = \varkappa E_\infty \tag{1.5a}$$

independently of d . One can use the mode-matching procedure known from the waveguide theory [3], expanding

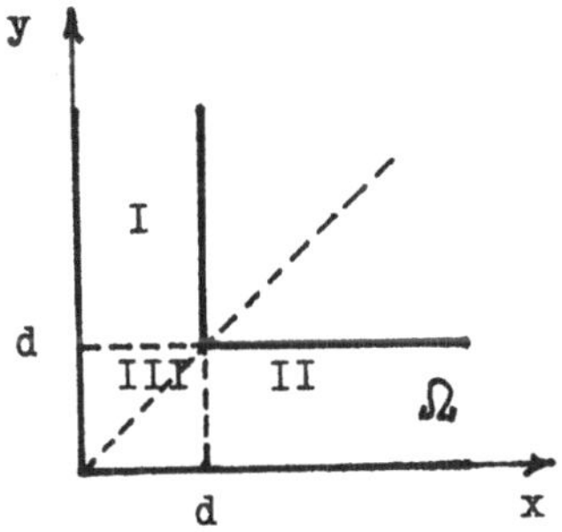

Fig.2

$$\psi_I = \sum_{j=1}^{\infty} (-1)^{j+1} r_j \, e^{q_j(1-y/d)} \varphi_j(x) \quad ,$$

where $q_j = \sqrt{j^2-E^2}$ and $\varphi_j = \sqrt{2/d}\,\sin(\pi jx/d)$, and similarly for the regions II and III . The regularity requirement then yields the system of equations

$$r = Cr \tag{1.6a}$$

for $r = \{r_j\}$, where

$$C_{jk} = \frac{1}{\pi}(1 - e^{-2q_j}) \frac{jk}{\sqrt{j^2-\varkappa}\,(j^2+k^2-\varkappa)} \tag{1.6b}$$

This system cannot be solved directly, because C is <u>not</u> compact on $l^2(j^{-1})$. It can be reformulated, however, into another system which is solvable and yields[4]

$$\varkappa = 0.93 \tag{1.5b}$$

Let us comment briefly on physical applications of these results. The most interesting among them is a possible existence of <u>edge-confined currents</u> in thin films grown on a sharply edged surface. The spectrum of the three-dimensional problem is, of course, continuous starting at E , however, the electrons with energy less than E_∞ cannot leave the edge. The transitions between E and the (two-dimensional) continuous spectrum are also observable in principle, being manifested by infrared photons.

2. The long-wave approximation in Q-guides

It is extremely difficult to get exact solutions describing particles in Q-guides. At the same time, it might not be necessary in some cases. Recall how useful are the point-interaction method in low-energy processes when the particle "sees" a point instead of an interaction potential which can be of a complicated structure but localized to a region small comparing to the Broglie wavelength[5].

One can try to use the same idea to describe the Q-guides replacing them by appropriate graphs (Fig.3). The question is whether it will work. Let us start with the curvature effects considered above.
We denote by E_d and $E_{\infty,d}$ the bound-state and first transversal-mode energies for a given width d , respectively. Let further E be

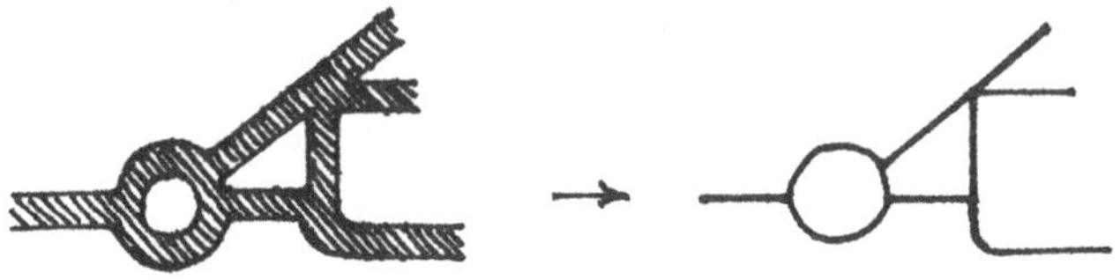

Fig.3

the ground state energy of the operator

$$H : \quad H\psi = \frac{\hbar^2}{2m}\left(-\frac{d^2}{ds^2} - \frac{1}{4}\gamma(s)^2\right)\psi \tag{2.1}$$

on $L^2(\mathbb{R})$ with the natural domain. A straightforward estimation procedure than gives

2.1 Theorem : Assume (a)-(e), then $\lim_{d\to 0+} (E_{\infty,d} - E_d) = E$.

This is the rigorous version of heuristic conclusions obtained by several authors [6-9].

The behaviour at branching points represents a more difficult problem. The simplest case is represented by the so-caled Y-junction (Fig.4). For the symmetric Y-junction ($\alpha = 2(\pi-\beta)$) and Dirichlet boundary conditions, the electromagnetic reflection and transmission coefficients $|s_{11}|$ and $|s_{12}|$ have been calculated in Ref.10 . A remarkable fact is that their zero-frequency limits are independent of α and equal to

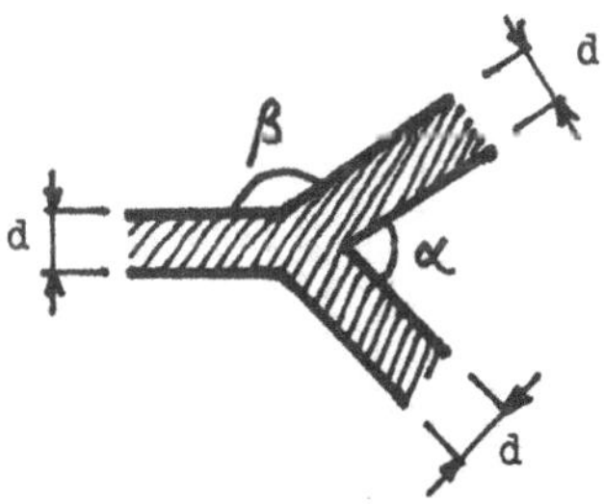

Fig.4

$$\lim_{\omega\to 0} |s_{11}| = \frac{1}{3} \quad , \quad \lim_{\omega\to 0} |s_{12}| = \frac{2}{3} \quad . \tag{2.2}$$

Let us mention further that for Y-junctions with a smooth boundary and Dirichlet b.c., existence of bound states can be proven in a similar way as in the previous section (there is no bound state in the Neumann case as it is well known from the classical waveguide theory).

Now one has to construct a suitable graph model for the Y-junction. It has been done in Ref.11 ; we review here the results. The appropriate graph consists of three halflines connected at one point ; the corresponding state Hilbert space is $\mathcal{H} = L^2(\mathbb{R}^+) \oplus L^2(\mathbb{R}^+) \oplus L^2(\mathbb{R}^+)$. The construction of Hamiltonian starts from the operator

$$H_0 = H_{0,1} \oplus H_{0,2} \oplus H_{0,3} \quad , \tag{2.3}$$

where $H_{0,j} = -(\hbar^2/2m)\, d^2/dx^2$ defined on $D(H_{0,j}) = C_0^{\infty}(\mathbb{R}^+)$. The operator (2.3) has deficiency indices (3,3) and hence a 9-parameter family of self-adjoint extensions. We restrict our attention to those among them that commute with the operators P_{jk} permuting the halflines. This leaves us with the 2-parameter family whose elements are specified uniquely by the boundary conditions

$$\begin{aligned} f_1'(0) &= Af_1(0) + Bf_2(0) + Bf_3(0) \quad , \\ f_2'(0) &= Bf_1(0) + Af_2(0) + Bf_3(0) \quad , \\ f_3'(0) &= Bf_1(0) + Bf_2(0) + Af_3(0) \end{aligned} \tag{2.4}$$

with $A,B \in \mathbb{R}$ and two "exceptional" one-parameter classes specified by the conditions

$$\begin{aligned} f_1'(0) + f_2'(0) + f_3'(0) &= 0 \quad , \\ f_2(0) - f_3(0) &= C(f_3'(0) - f_2'(0)) \quad , \\ f_1(0) - f_3(0) &= C(f_3'(0) - f_1'(0)) \end{aligned} \tag{2.5}$$

and

$$\begin{aligned} f_1'(0) = f_2'(0) = f_3'(0) &\equiv f'(0) \quad , \\ f_1(0) + f_2(0) + f_3(0) &= Df'(0) \quad , \end{aligned} \tag{2.6}$$

where again $C,D \in \mathbb{R}$. The S-matrices for scattering on the junction corresponding to each of these Hamiltonians are easily calculated to be

$$S(k) = \begin{pmatrix} a & b & b \\ b & a & b \\ b & b & a \end{pmatrix} \tag{2.7}$$

with

$$a = \frac{-1 + ikB - k^2(A^2+AB-2B^2)}{[1 - ik(A+2B)][1 - ik(A-B)]}$$

$$b = \frac{-2ikB}{[1 - ik(A+2B)][1 - ik(A-B)]}$$

for the boundary conditions (2.4), and

$$S(k) = \frac{1}{3(1+ikC)} \begin{pmatrix} 3ikC-1 & 2 & 2 \\ 2 & 3ikC-1 & 2 \\ 2 & 2 & 3ikC-1 \end{pmatrix} \qquad (2.8)$$

$$S(k) = \frac{1}{3-ikD} \begin{pmatrix} 1-ikD & -2 & -2 \\ -2 & 1-ikD & -2 \\ -2 & -2 & 1-ikD \end{pmatrix} \qquad (2.9)$$

for the remaining cases.

What is the physical meaning of the parameters appearing in the conditions (2.4)-(2.6) ? A natural conjecture is that they are related to the angles α, β specifying a given Y-junction. In particular, we conjecture that to a symmetric Y-junction one of the "exceptional" Hamiltonians corresponds. This guess is supported by the following argument : the S-matrices (2.8) and (2.9) satisfy

$$\lim_{k \to 0+} S(k) = \pm \begin{pmatrix} -1/3 & 2/3 & 2/3 \\ 2/3 & -1/3 & 2/3 \\ 2/3 & 2/3 & -1/3 \end{pmatrix}$$

independently of the parameters C and D, which is just the feature expressed by the relation (2.2) . Of course, the quantum-mechanical reflection and transmission coefficients are given by squared moduli of the S-matrix elements, being therefore equal to 1/9 and 4/9, respectively. Furthermore, we see that a part of the Hamiltonians (one half of them, roughly speaking) has a bound state corresponding to a pole in the S-matrix. It suggests that just this part describes the Y-junctions with Dirichlet boundary conditions.

Hence the long-wave approximation described above represents a useful way in which one can replace the original problem by a mathematically much more simple one. Among its potential applications, those related to Aharonov-Bohm-type effects in microstructures [12-14] are particularly interesting. We limit ourselves here to one of them, namely to a description of a semiconductor loop with two external leads in a homogeneous electric field (Fig.5). In order to calculate the transmission coefficient through the loop, one has

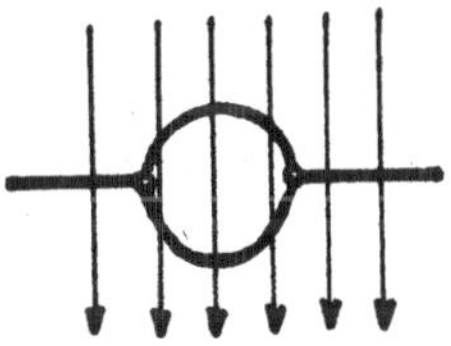

Fig.5

to "sew" using the boundary conditions (2.4) the wavefunctions

$$u_1(x) = e^{-ikx} + a\, e^{ikx} \quad , \quad u_4(x) = b\, e^{-ikx}$$

describing the electron behaviour on the leads, and the functions u_2, u_3 which solve the equations

$$-\frac{\hbar^2}{2m} u_j''(x) + V_j(x) u_j(x) = E u_j(x)$$

on the arms of the loop. Here $E = \hbar^2 k^2/2m$ and the potentials V_j are given by the electric field (they should contain also a curvature-dependent term - cf.(2.1) - but it is negligible in the physically interesting situations).

The problem leads to a system of linear equations which yields the following expressions for the transmission coefficient [15,16]

$$T(E) = |b|^2 = \frac{B_2^2}{1 + k^2 A_2^2} |c_2 + d_2|^2 \quad , \tag{2.10}$$

where

$$\begin{pmatrix} c_1 \\ c_2 \end{pmatrix} = C_2(-k) \begin{pmatrix} d_1 \\ d_2 \end{pmatrix} \quad ,$$

$$\begin{pmatrix} d_1 \\ d_2 \end{pmatrix} = \left[\Pi_2^{-1} C_2(-k) - C_1(-k) \Pi_3^{-1} \right]^{-1} \begin{pmatrix} z_1 \\ z_2 \end{pmatrix}$$

with

$$z_1 = (A_1 - B_1) z_2 \quad , \quad z_2 = \frac{2ik}{1 + ik(B_1 - A_1)} \quad ,$$

$$C_1(k) =$$

$$\frac{1}{B_1 \left[1 + ik(B_1 - A_1) \right]} \begin{pmatrix} A_1 + ik(B_1^2 - A_1^2) & (B_1 - A_1)\left[(A_1 + B_1)(1 - ikA_1) + 2ikB_1^2 \right] \\ 1 - ikA_1 & -A_1 - ik(B_1^2 - A_1^2) \end{pmatrix}$$

and $C_2(k)$ expresses similarly through the parameters A_2, B_2 specifying the second junction. Furthermore, Π_j are the transfer matrices

$$\begin{pmatrix} u_j(l_j) \\ u_j'(l_j) \end{pmatrix} = \Pi_j \begin{pmatrix} u_j(0) \\ u_j'(0) \end{pmatrix} \quad , \quad j = 2,3 \quad .$$

Transport properties of the loop are then given by Landauer formula [17] : the conductance equals

$$G = \frac{e^2}{\pi\hbar} \frac{T(E)}{1 - T(E)} \tag{2.11}$$

recall that $\pi\hbar/e^2 \approx 12906\,\Omega$) . Using the WKB-expressions for Π_j , one can calculate G from (2.10) and (2.11) numerically for various shapes of the loop and parameters specifying the junctions. As an example, we present here (Fig.6) the results for a loop of a 200 Å GaAs wire of the sketched shape with $A_1 = B_1 = -A_2 = -B_2 = 1$. We see that the conductance plot exhibits sharp minima at reasonably low field intensities. Changing the loop shape and parameters, we obtain similar pictures. This interference effect is particularly promising from the viewpoint of possible device applications.

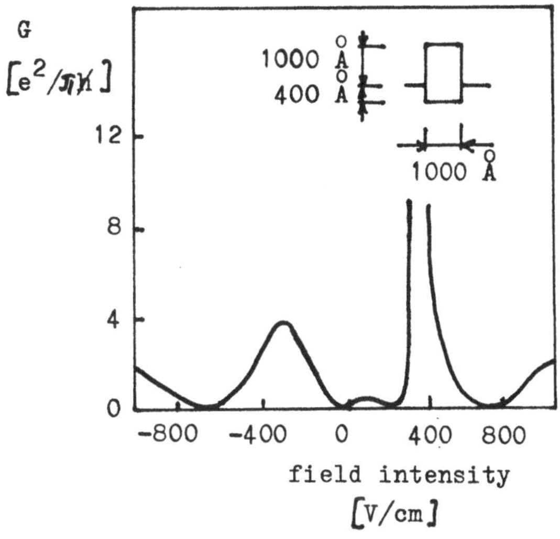

Fig.6

In conclusion, let us mention that the long-wave approximation can be used effectively in other fields too, e.g., in the electromagnetic waveguide theory, for killing the humming noise in air-conditioning systems etc.

References

1 H.Sakaki, in Proc.Symp.Foundations of Quantum Mechanics, Phys. Soc.of Japan, Tokyo 1983 ; pp.94-110.

2 P.Exner, P.Šeba : Bound states in curved quantum waveguides, preprint, Bochum 1987.

3 R.Mittra, S.W.Lee : Analytical Techniques in the Theory of Guided Waves, Macmillan, New York 1971.

4 P.Exner, P.Šeba, P.Šťovíček : A correct truncation procedure for a rectangular planar waveguide, in preparation.

5 S.Albeverio, F.Gesztesy, R.Høegh-Krohn, H.Holden : Solvable Models in Quantum Mechanics, Springer, Berlin 1988.

6 J.Marcus, J.Chem.Phys.45 (1966),4493-4499.

7 H.Jensen, H.Koppe, Ann.Phys.63 (1971),127-140.

8 D.Picca, Lett.N.Cim.34 (1982),449-452.

9 R.C.T.da Costa, Phys.Rev.A23 (1981),1982-1987 ; A25 (1982), 2893-2900.

10 R.Mehran, IEEE Trans.26 (1978), 400-405.

11 P.Exner, P.Šeba : Free quantum motion on a branching graph I,II , preprints JINR E2-87-213,214, Dubna 1987.

12 R.A.Webb et al., Phys.Rev.Lett.54 (1985),2696-2700.

13 C.P.Umbach et al., Appl.Phys.Lett.50 (1987),1289-1291.

14 C.P.Umbach et al., Phys.Rev.Lett.56 (1986),386-389.

15 P.Exner, P.Šeba : On feasibility of quantum interference transistors, preprint JINR E2-87-346, Dubna 1987.

16 P.Exner, P.Šeba, P.Šťovíček : Quantum interference on graphs controlled by an external electric fiels, preprint JINR E2-87-707, Dubna 1987.

17 R.Landauer, Phys.Lett.A85 (1981),91-93.

AN EXACTLY SOLVABLE MODEL OF A CRYSTAL WITH NON-POINT ATOMS

Yu.A.Kuperin, K.A.Makarov, B.S.Pavlov
Department of Mathematical and Computational Physics, Institute for Physics, Leningrad State University, 198904 Leningrad, St.Peterhoff, USSR

1. Introduction

The Schrödinger equation with a periodic potential in R^3 provides the conventional one-electron solid-state model of a hard lattice. This model is exactly solvable in the zero-range point interactions approximation (the so-called Kronig-Penney model). The corresponding spectral picture seems to be rather poor. The reason is that the traditional Fermi deuteron potential which describes an isolated atom [1-4] has very simple spectral properties. The next step is to provide the individual atoms with an internal structure [5] using the method of the Hilbert-space extensions and to study properties of the corresponding crystal lattice. In such models one can obtain an information about the influence of the sharp atomic resonances on the band structure of the crystal and get answers to some other general questions ([6],[7]).

The shortcommings of the models described above are that they can describe the effects of s-wave scattering only, because the higher angular momenta can not be included into the zero-range approximation in a mathematically well - defined way (see [8]). This difficulty can be overcome in the models of atoms with a finite size core such as bag-like models [9] and their modifications [10,11] in which the presence of the atomic internal structure is imitated by an abstract Hamiltónian.

In this paper we consider a three-dimensional cubic lattice in R^3 . The individual atoms which form the lattice are supposed in our model to be semitransparent closed surfaces equiped with an internal structure. The "centres" of these surfaces are located on the lattice sites $n \in Z^3$. The construction of the Bloch waves in this model can be reduced to solution of a Schroedinger equation with special energy-dependent boundary conditions on the interaction surfaces. The boundary conditions are defined by the internal channel Hamiltònian of the scatterer and are responsible for both the spectral properties of an isolated atom and the band structure of the crystal. The aim of the present paper is to derive the dispersion relations connecting the energy and the quasimomentum of Bloch waves in this model.

2. The Hamiltonian of an individual atom

The Hamiltonian of the isolated atom with an internal structure acts in the direct sum of the space $L_2(R^3)$ describing the external degrees of freedom and some abstract Hilbert space H of internal degrees of freedom. The unperturbed dynamics in the external channel is determined by a Laplacian, while the internal - channel dynamics is governed by an abstract bounded self-adjoint operator A. We consider here the simplest case of coupling between the external and internal channels setting suitable boundary conditions on a closed surface S_0 in the exterior configuration space.

The Hamiltonian for the individual atom can be constructed from the unperturbed operator $-\Delta \oplus A$ in the following manner. We restrict the external part $\Delta \to \Delta_0$ to the linear set of all smooth functions vanishing together with their normal derivatives near the interaction surface S_0. Then we restrict the internal component of the operator A to A_0 in such a way that a fixed vector $e \in H$ becomes a deficiency element of A_0 : $A_0 e = i e$ [5]. Then the domain $D(A_0^*)$ of the adjoint operator A_0^* can be described in terms of the so-called "real" basis $w^+ = A(A-iI)^{-1}e$, $w^- = (A-iI)^{-1}e$ as follows

$$u_1 = \tilde{u}_1 + \varepsilon^+ w^+ + \varepsilon^- w^- , \quad u_1 \in D(A_0^*) \tag{1}$$

where

$$\tilde{u}_1 = (A-iI)^{-1}\psi , \psi \perp e \tag{2}$$

The total boundary form of the operator $h_0^* = (-\Delta_0)^* \oplus A_0^*$ restricted to domain $U = \{(u_0(x), u_1) : u_0(x)$ is a smoth function in the vicinity of S_0 , $u_1 \in H\}$ can be written as

$$\langle h_0^* U, V\rangle - \langle U, h_0^* V\rangle = \int_{S_0} ([n \nabla u_0]\overline{v_0} - u_0 \overline{[n \nabla v_0]}) ds +$$

$$+ \varepsilon^-(U)\overline{\varepsilon^+(V)} - \varepsilon^+(U)\overline{\varepsilon^-(V)} \tag{3}$$

where $[n\nabla u] = n\nabla u|_{S_0^+} - n\nabla u|_{S_0^-}$, and $\mathcal{E}^{\pm}$ are boundary values of the element u_1 [5]. The Hamiltonian h of the individual atom is non-trivial self-adjoint extension of the operator h_0. Such an extension can be described in terms of boundary conditions under which the symplectic form (3) vanishes, for example,

$$\begin{pmatrix} [n\nabla u_0] \\ \mathcal{E}^- \end{pmatrix} = B \begin{pmatrix} u_0 \\ \mathcal{E}^+ \end{pmatrix} \tag{4}$$

Here the self-adjoint operator B has the following matrix representation

$$B = \begin{pmatrix} \gamma_e & \varphi \\ \langle\ ,\varphi\rangle & \gamma_i \end{pmatrix},$$

where $\gamma_e, \gamma_i \in R, \varphi \in L_2(S_0)$ are parametres of the model and the function φ generates the functional

$$\langle u_0, \varphi\rangle = \int_{S_0} u_0 \bar{\varphi}\, ds$$

The spectral analysis of the Hamiltonian h has been performed in Ref. 12. In particular, the external components u_0 of the eigenfunctions U : $hU = \lambda U$, are known to be the solutions of the following energy-dependent boundary value problem

$$-\Delta u_0 = \lambda u_0, \tag{5}$$

$$[n\nabla u_0] = \gamma_e u_0 + (D(\lambda) - \gamma_i)^{-1} \langle u_0, \varphi\rangle \varphi$$

The boundary conditions in (5) follows from (4) combined with the the equality $\mathcal{E}^- = D(\lambda)\mathcal{E}^+$, where $D(\lambda)$ is the Schwarz integral [5] of the self-adjoint operator A :

$$D(\lambda) = \langle (I + \lambda A)(A - \lambda I)^{-1} \theta, \theta\rangle$$

The internal components u_1 of the wave functions U can be reconstructed from the external ones. Thus it is sufficient to study the boundary value problem (5) only.

The external components of the eigenfunctions belonging to the point spectrum of the Hamiltinian h are the solutions of the problem (5). They can be sought in the form of simple-layer potentials

$$u_0(x) = \int_{y \in S_0} G(x-y, \sqrt{\lambda})\, \rho(y)\, dy \equiv G\rho \tag{6}$$

Here $$G(x-y, \sqrt{\lambda}) = \frac{1}{4\pi} \frac{\exp i\sqrt{\lambda}|x-y|}{|x-y|}$$

is the Green's function of the Laplacian and $\sqrt{\lambda}$ is choosen in such a way that $\mathrm{Im}\sqrt{\lambda} > 0$.

To fulfill (5) the simple-layer density ρ must be a solution of the integral equation

$$\rho = \gamma_e G\rho + (D(\lambda) - \gamma_i)^{-1} \langle G\rho, \varphi\rangle \varphi \qquad (7)$$

The point spectrum of h consists of negative λ for which the eq(7) has a non-trivial smooth solution. The eigenfunctions of the absolutely continuous spectrum of the Hamiltonian h can be found by the same technique. For this purpose it is sufficient to add an incoming plane wave $\exp i(k,x)$ with the wave vector k, $k^2 = \lambda$, to the simple-layer potential (6). Then the corresponding density ρ can be found as a solution to the inhomogeneous Fredholm equation associated with (7).

3. The lattice Hamiltonian

The unperturbed lattice Hamiltonian $-\Delta \oplus \sum^{\oplus}_{n \in Z^3} A_n$ acts in the direct sum of spaces $L_2(R^3)$ and H_n, where $A_n = A$ act in the identical copies H_n of the Hilbert space H. Let us restrict the Laplacian to the linear set of functions vanishing in a neighbourhood of the surfaces S_n, which are supposed to be copies of S_0 placed on sites of the lattice and small enough to avoid intersecting. Then we restrict the operators A_n to symmetric operators using the procedure described above. The total Hamiltonian is defined as a self-adjoint extension of the corresponding restricted operator defined on the linear set of funtions satisfying the boundary conditions (4) on every S_n. In this paper, we are going to study the external components of Bloch waves and to derive the dispersion relations.

The external components of Bloch waves can be found as a solution of the equation $-\Delta u = k^2 u$ satisfying on the surfaces S_n the following boundary conditions

$$[n \nabla u]\Big|_{S_n} = \gamma_e u\Big|_{S_n} + (D(\lambda) - \gamma_i)^{-1} \langle u, \varphi \rangle_{L_2(S_n)}, \qquad (8)$$

where we have used the Bloch's anzatz

$$u(x+n) = \exp i(t,n)\, u(x), \qquad n \in Z^3 \qquad (9)$$

and t is the quasimomentum of the system. Such solutions can be represented as s sum of simple-layer potentials corresponding to the surfaces S_n centred at the lattice sites $n \in Z^3$:

$$u(x) = \sum_{n \in Z^3} \int_{y \in S_n} G(x-y, \sqrt{\lambda})\, \rho_n(y)\, dy \qquad (10)$$

with the densities ρ_n satisfying on S_n the Bloch's boundary conditions

$$\rho_n(x+y) = \rho_o(x) \exp i(t,n) , \quad x \in S_o \qquad (11)$$

To fulfil (8) the densities ρ_n must be related mutually by

$$\rho_m(x) = \left[\gamma_e + \frac{\langle \cdot , \varphi \rangle \varphi}{D(\lambda) - \gamma_i} \right] \sum_{n \in Z^3} \int_{y \in S_n} G(x-y, \sqrt{\lambda})\, \rho_n(y) dy \qquad (12)$$

The equations (12) together with the Bloch conditions (11) give

$$\rho_o(t,x) = \left[\gamma_e + \frac{\langle \ , \varphi \rangle \varphi}{D(\lambda) - \gamma_i} \right] \sum_{n \in Z^3} \int_{y \in S_o} G(x-y+n, \sqrt{\lambda}) \rho_o(t,y) e^{itn}, \quad x \in S_o \qquad (13)$$

Introducing the lattice sum

$$K(x-y, t, \sqrt{\lambda}) = \sum_{n \in Z^3} \frac{1}{4\pi} \frac{\exp i\sqrt{\lambda}|x-y-n|}{|x-y-n|} \exp i(t,n) \qquad (14)$$

one can rewrite eq. (13) as an integral equation with the kernel K :

$$\rho_o(t,x) = \left[\gamma_e + \frac{\langle \ , \varphi \rangle \varphi}{D(\lambda) - \gamma_i} \right] \int_{y \in S_o} K(x-y, t, \sqrt{\lambda})\, \rho_o(t,y)\, dy \qquad (15)$$

Inverting the operator which appears in the relation (15),

$$\left[\gamma_e I + \frac{\langle \cdot , \varphi \rangle \varphi}{D(\lambda) - \gamma_i} \right]^{-1} = \gamma_e^{-1} \quad I - \frac{\langle \cdot , \varphi \rangle \varphi}{|\varphi|^2 + \gamma_e (D(\lambda) - \gamma_i)} \qquad (16)$$

we get

$$\rho_o - \gamma_e K \rho_o = \frac{\langle \rho_o , \varphi \rangle \varphi}{|\varphi|^2 + \gamma_e (D(\lambda) - \gamma_i)} \qquad (17)$$

If the inverse operator $(I - \gamma_e K)^{-1}$ exists (this is true, e.g., in the weak-coupling case $\gamma_e \ll 1$) , we denote $R = (I - \gamma_e K)^{-1} - I$. The eq. (17) in terms of this operator reads

$$\rho_0(t,x) = \frac{\langle \rho_0, \varphi \rangle}{|\varphi|^2 + \gamma_e(D(\lambda) - \gamma_i)}(\varphi + \int_{y\in S_0} R(x-y,t,\sqrt{\lambda})\varphi(y)\, dy \quad (18)$$

and projecting the last equation on $\varphi \in L_2(S_0)$, we get a homogeneous algebraic equation. The conditions of its solvability yields the following dispersion relation

$$\gamma_e(D(\lambda) - \gamma_i) = \langle R(t,\sqrt{\lambda})\varphi, \varphi \rangle_{L_2(S_0)} \quad (19)$$

If $\gamma_e = 0$ this relation must be modified. To this end, let us return to the eq.(18), which is solvable for $\gamma_e = 0$ if

$$D(\lambda) - \gamma_i = \langle K(t,\sqrt{\lambda})\varphi, \varphi \rangle_{L_2(S_0)} \quad (20)$$

which is the sought dispersion relation in the $\gamma_e = 0$ case.

In conclusion, let us note that the role of the projection $\langle \ , \varphi \rangle \varphi$ in the boundary conditions (8) can be played by an arbitrary self-adjoint operator in $L_2(S_0)$ or even by an operator-valued R-function (13). This enriches the scattering matrix of the individual atom, and therefore also spectral properties of the whole crystal would be more complicated. In the model we have proposed here all the partial waves can contribute to the scattering processes of electrons on the crystal. This makes it possible to model properties of real crystals outside the perturbation-theory framework.

Finally, the model can be easily generalized to crystal lattices without the cubic symmetry. Moreover, it can describe crystals containing different kinds of atoms. The only modification required in these casis is the refinement of the lattice sum.

References

1 Yu.N.Demkov, V.N.Ostrovskii: The Zero-Range Potentials Approach to Nuclear Physics, Leningrad State Univ. 1975(in Russian).
2 A.Grossmann,R.Høegh-Krohn, M.Mebkhout, J.Math.Phys. 21(1980),2376
3 Yu.E.Karpeshina, Teor. Mat. Fiz. 57(1983), 304
4 E.Fermi, Nuovo Cimento, 11(1934),157
5 B.S.Pavlov, Teor. Mat. Fiz 59(1984),345
6 B.S.Pavlov, N.V.Smirnov, in "Probl. Mat. Fiz." , Leningrad State Univ. 1987, v.12, p.155
7 B.S.Pavlov, S.E.Cheremshantsev, Preprint LOMI P-15-85, Leningrad 1985

8 F.I.Dalidtchik, V.Z.Slonim, JETP Letters 31(1980), 122
9 A.W.Thomas, Adv. Nucl. Phys., 13(1984), 1
10 Yu.A.Kuperin, K.A.Makarov, B.S.Pavlov, Teor. Mat.Fiz.69(1986),100
11 Yu.A.Kuperin, K.A.Makarov, S.P.Merkuriev, B.S.Pavlov, A.K.Motovilov, ITP-Budapest-Report N 441, Budapest 1986
12 Yu.A.Kuperin, K.A.Makarov, S.P.Merkuriev, A.K.Motovilov, B.S.Pavlov, Proc. of the School on Few-Body Quark-Hadronic Systems, Vilnus 1986, p.28
13 F.V.Atkinson : Discrete and Continuous Boundary Problems, Academic Press, New York 1964